At Sylvan, we believe that a lifelong love of learning begins at an early age, and we are glad you have chosen our resources to help your children experience the joy of mathematics as they build critical reasoning skills. We know that the time you spend with your children reinforcing the lessons learned in school will contribute to their love of learning.

Success in math requires more than just memorizing basic facts and algorithms; it also requires children to make sense of size, shape, and numbers as they appear in the world. Children who can connect their understanding of math to the world around them will be ready for the challenges of mathematics as they advance to more complex topics.

We use a research-based, step-by-step process in teaching math at Sylvan that includes thought-provoking math problems and activities. As students increase their success as problem solvers, they become more confident. With increasing confidence, students build even more success. The design of the Sylvan workbooks will help you to help your children build the skills and confidence that will contribute to success in school.

Included with your purchase of this workbook is a coupon for a discount at a participating Sylvan center. We hope you will use this coupon to further your children's academic journeys. Let us partner with you to support the development of confident, well-prepared, independent learners.

The Sylvan Team

Sylvan Learning Center
Unleash your child's potential here

No matter how big or small the academic challenge, every child has the ability to learn. But sometimes children need help making it happen. Sylvan believes every child has the potential to do great things. And, we know better than anyone else how to tap into that academic potential so that a child's future really is full of possibilities. Sylvan Learning Center is the place where your child can build and master the learning skills needed to succeed and unlock the potential you know is there.

The proven, personalized approach of our in-center programs delivers unparalleled results that other supplemental education services simply can't match. Your child's achievements will be seen not only in test scores and report cards but outside the classroom as well. And when your child starts achieving his or her full potential, everyone will know it. You will see a new level of confidence come through in all of your child's activities and interactions.

How can Sylvan's personalized in-center approach help your child unleash the potential you know is there?

• Starting with our exclusive Sylvan Skills Assessment®, we pinpoint your child's exact academic needs.

• Then we develop a customized learning plan designed to meet your child's academic goals.

• Through our method of skill mastery, your child will not only learn and master every skill in a personalized plan, but he or she will be truly motivated and inspired to achieve.

To get started, included with this Sylvan product purchase is $10 off our exclusive Sylvan Skills Assessment®. Simply use this coupon and contact your local Sylvan Learning Center to set up your appointment.

To learn more about Sylvan and our innovative in-center programs, call 1-800-EDUCATE or visit www.SylvanLearning.com. *With over 900 locations in North America, there is a Sylvan Learning Center near you!*

2nd Grade
Geometry Success

Published in the United States by Random House, Inc., New York, and in Canada by Random House of Canada Limited, Toronto.

www.tutoring.sylvanlearning.com

Created by Smarterville Productions LLC
Producer & Editorial Direction: The Linguistic Edge
Producer: TJ Trochlil McGreevy
Writer: Amy Kraft
Cover and Interior Illustrations: Shawn Finley, Tim Goldman, and Duendes del Sur
Layout and Art Direction: SunDried Penguin

First Edition

ISBN: 978-0-307-47927-3
ISSN: 2156-6410

This book is available at special discounts for bulk purchases for sales promotions or premiums. For more information, write to Special Markets/Premium Sales, 1745 Broadway, MD 6-2, New York, New York 10019 or e-mail specialmarkets@randomhouse.com.

PRINTED IN CHINA

10 9 8 7 6 5 4 3 2 1

Contents

Identifying Plane Shapes

Color the Shapes

COLOR all of the shapes according to the color of the shapes at the top of the page.

HINT: A square is a special kind of rectangle.

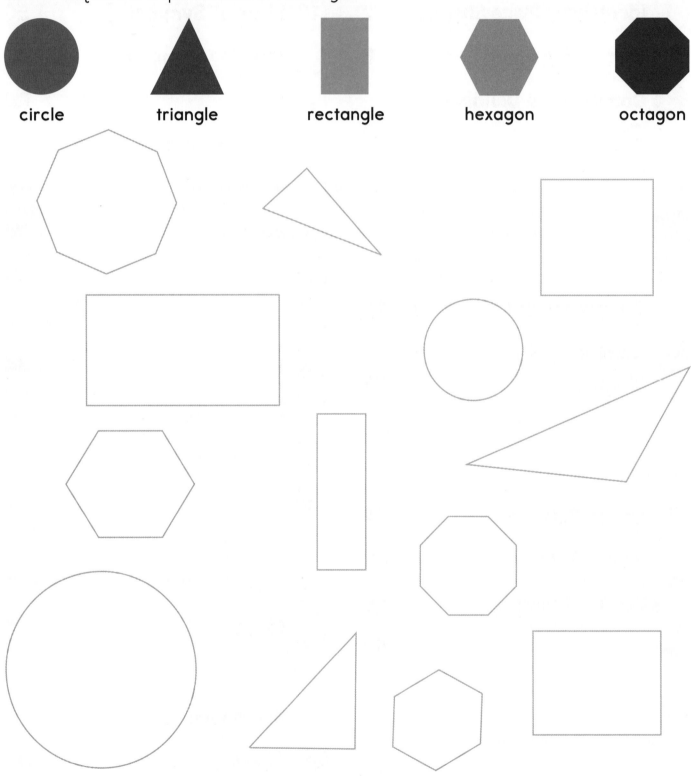

circle   triangle   rectangle   hexagon   octagon

Cross Out

CROSS OUT any picture that is **not** a rectangle.

HINT: A square is a special kind of rectangle.

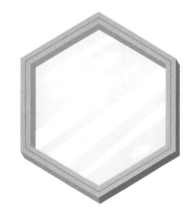

Shape Counter

WRITE the number of times each kind of shape appears on this page.

HINT: A square is a special kind of rectangle.

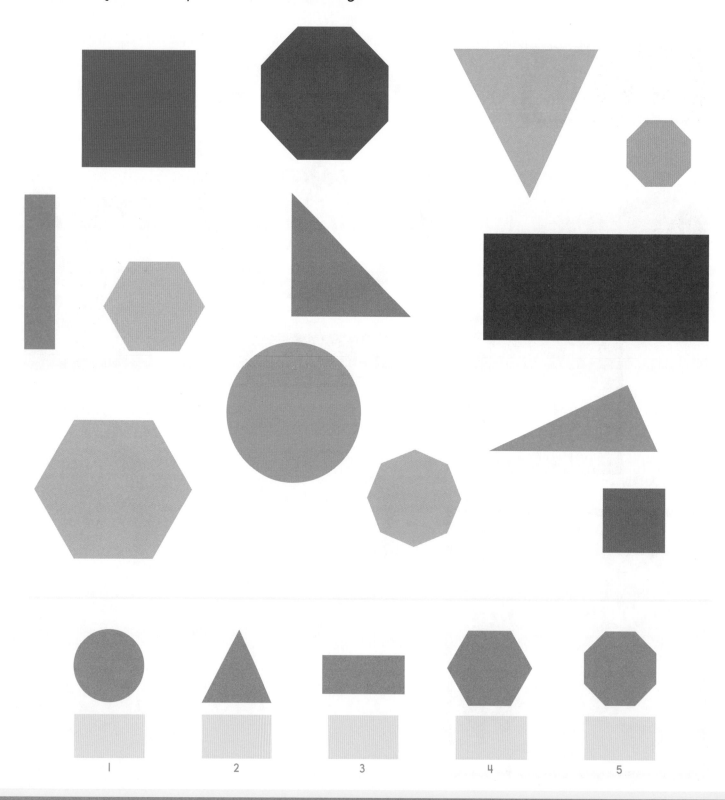

| 1 | 2 | 3 | 4 | 5 |

House Hunt

WALK around your house or yard to find things that are the same shape as these shapes. WRITE what you find.

Match Up

DRAW lines to connect the shapes that are the same.

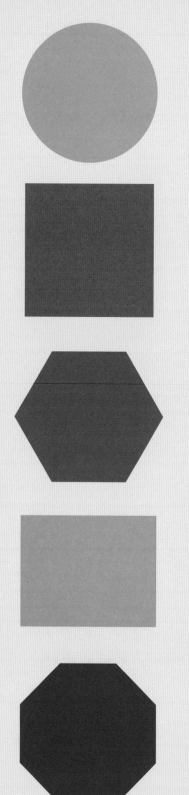

Which One?

CIRCLE the picture in each row that is the same shape as the red shape.

Comparing Plane Shapes

Which Window?

DRAW a line from each window to the hole in the wall with the same shape.

Shape Up

DRAW four different triangles and four different rectangles.

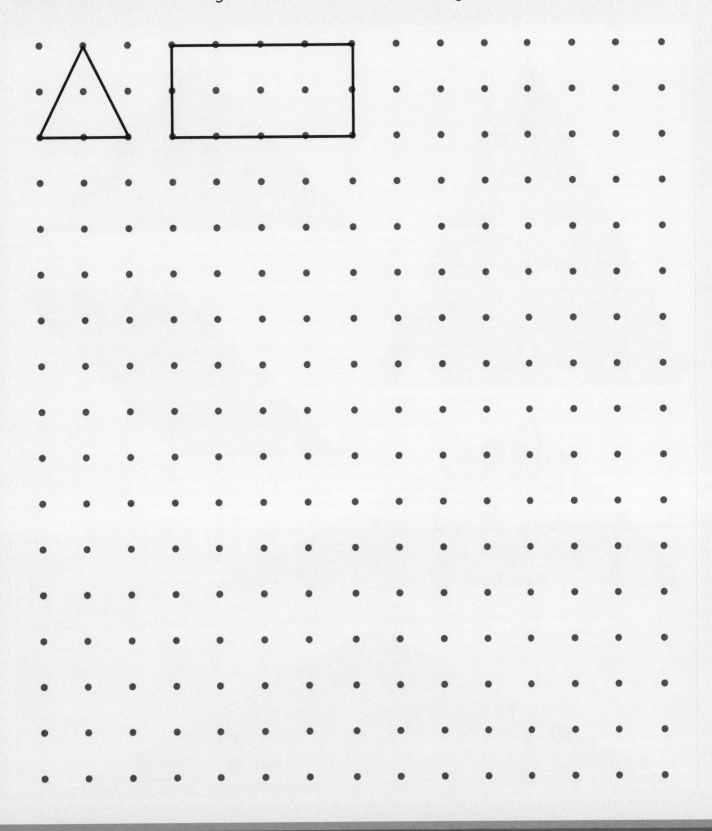

Shape Puzzler

CUT OUT the triangles on page 11, and ARRANGE them to make each shape shown.
(Save the pieces to use later in the workbook.)

Shifting Shapes

WRITE the number of triangles, rectangles, and squares you see.

HINT: Think about the different ways smaller shapes can make larger shapes.

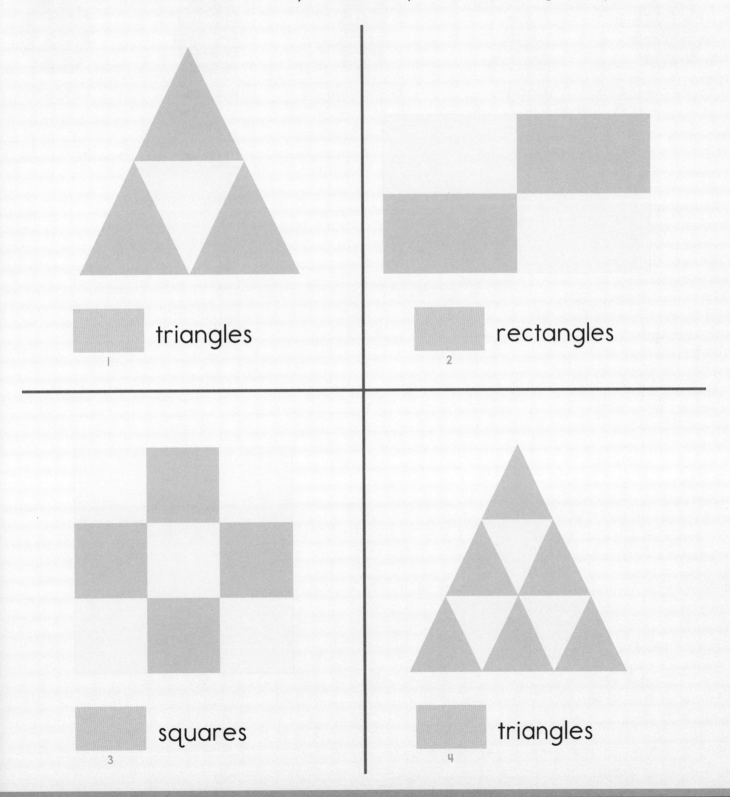

triangles

1

rectangles

2

squares

3

triangles

4

Color the Shapes

COLOR all of the shapes according to the color of the shapes at the top of the page.

circle

triangle

rectangle

hexagon

octagon

Robot Repair

DRAW a line from each robot head to the body with the matching shape.

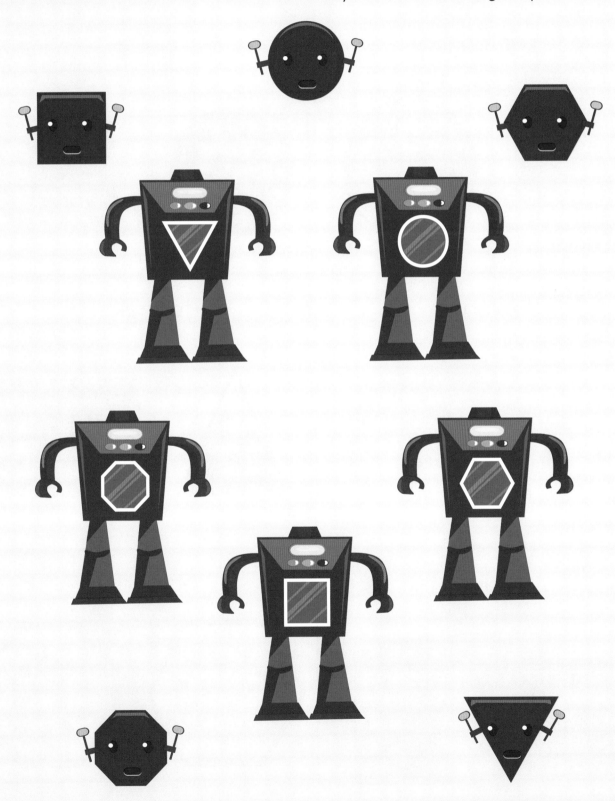

Review

Which One?

CIRCLE the correct picture in each pair.

rectangle

circle

 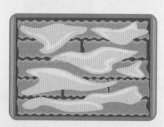

hexagon

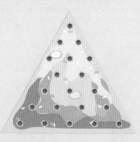 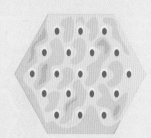

triangle

square

 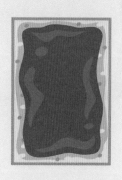

octagon

Shape Puzzler

Using the triangles from page II, ARRANGE them to make each shape shown.

Identifying Solid Shapes

Name That Shape

WRITE the name of each shape.

| sphere | cylinder | rectangular prism | cube | square pyramid | cone |

1. _____

2. _____

3. _____

4. _____

5. _____

6. _____

7. _____

8. _____

9. _____

Cross Out

CROSS OUT any picture that is **not** a rectangular prism.

HINT: A cube is a special kind of rectangular prism.

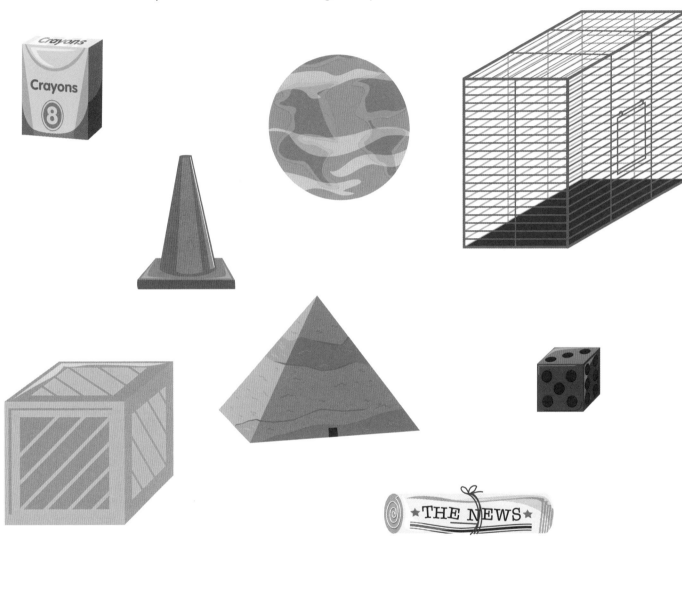

Identifying Solid Shapes

Shape Counter

WRITE the number of times each kind of shape appears on this page.

HINT: A cube is a special kind of rectangular prism.

1   2   3   4   5

House Hunt

WALK around your house or yard to find things that are the same shape as these shapes. WRITE what you find.

Shape Builders

CUT OUT each shape on the opposite page. FOLD on the dotted lines, and GLUE the tabs to construct each shape. Then WRITE the answers to the questions.

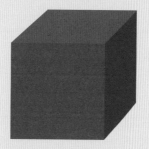

1. What is the name of the red shape?

_____

2. What is the name of the purple shape?

_____

3. How are these shapes alike?

_____

4. How are these shapes different?

_____

5. Name one thing that looks like the red shape.

_____

6. Name one thing that looks like the purple shape.

_____

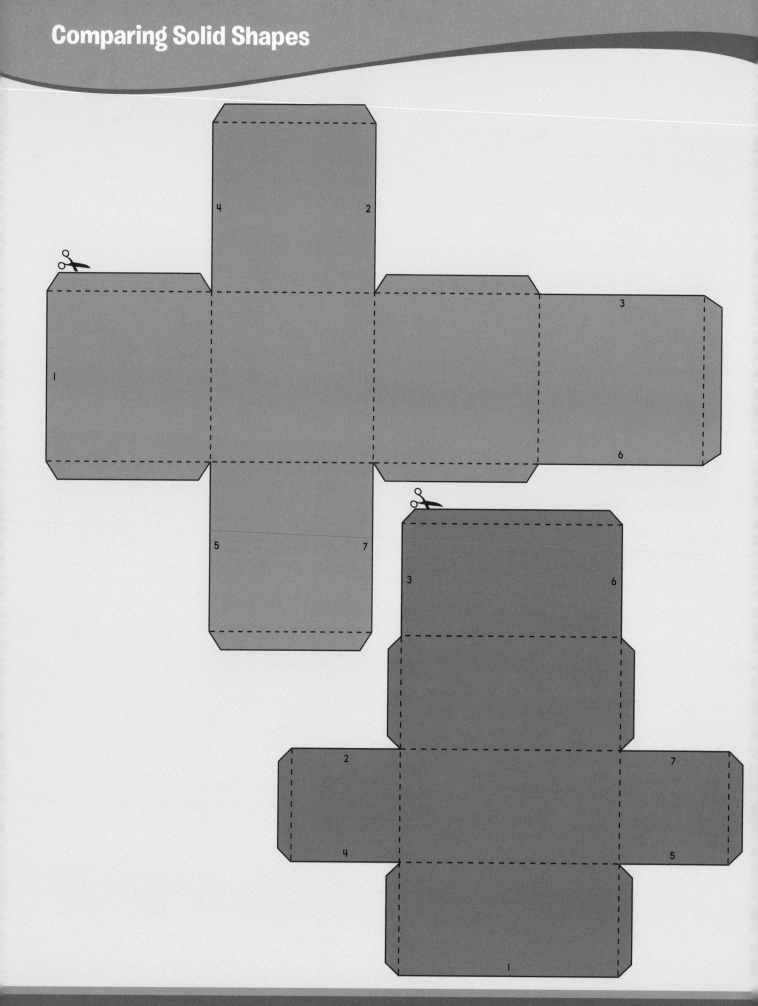

Find the Same

CIRCLE the picture in each row that is the same kind of shape as the first shape.

Combining & Dividing Shapes

Match Up

DRAW lines to connect the shapes with the larger shape they make when they are put together.

Match Up

DRAW lines to connect the shapes with the picture they make when they are put together.

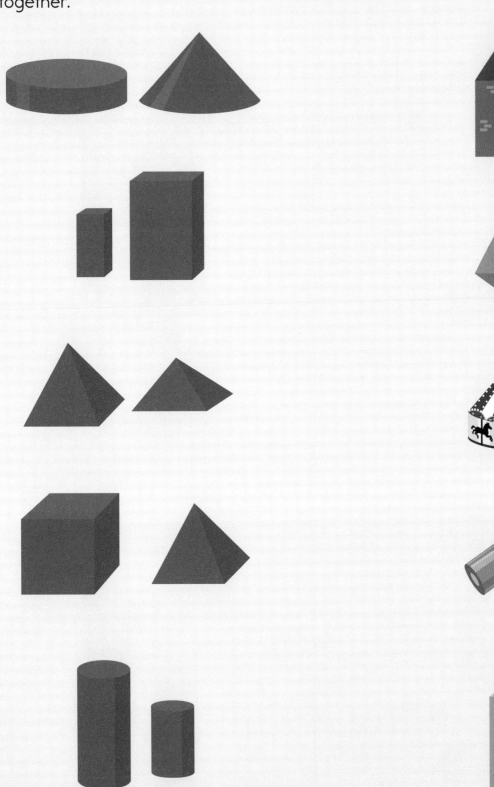

Combining & Dividing Shapes

Shape Slices

CIRCLE the two shapes in each set that can be combined to make the top shape.

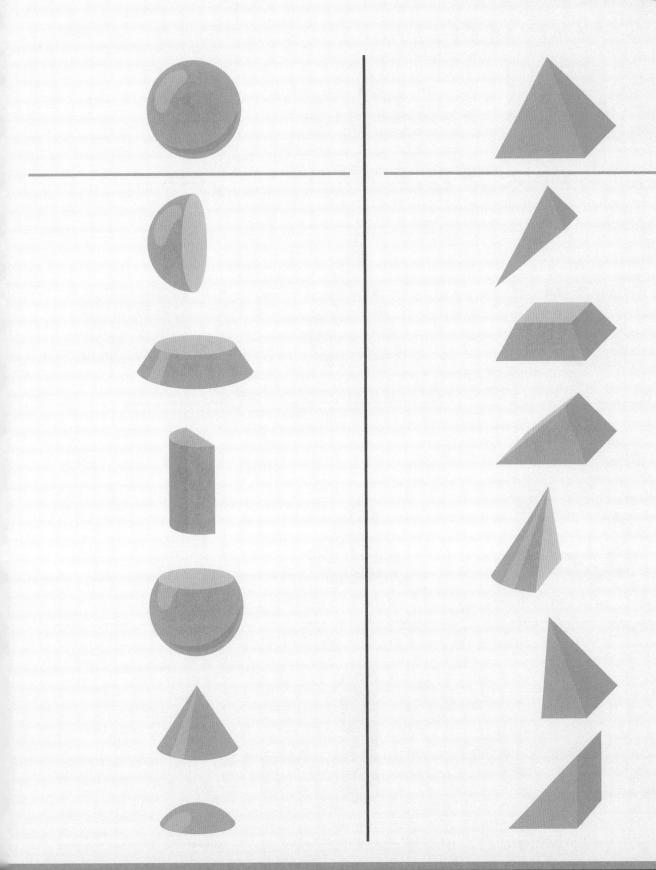

Name That Shape

WRITE the name of each shape.

1. _____

2. _____

3. _____

4. _____

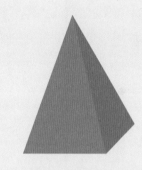

5. _____

6. _____

7. _____

8. _____

9. _____

10. _____

11. _____

12. _____

Which One?

CIRCLE the correct picture in each pair.

rectangular prism

cylinder

square pyramid

cone

sphere

cube

Odd One Out

CROSS OUT the shape in each row that does **not** have the same kind of shape as the others.

Shape Slices

CIRCLE the two shapes in each set that can be combined to make the top shape.

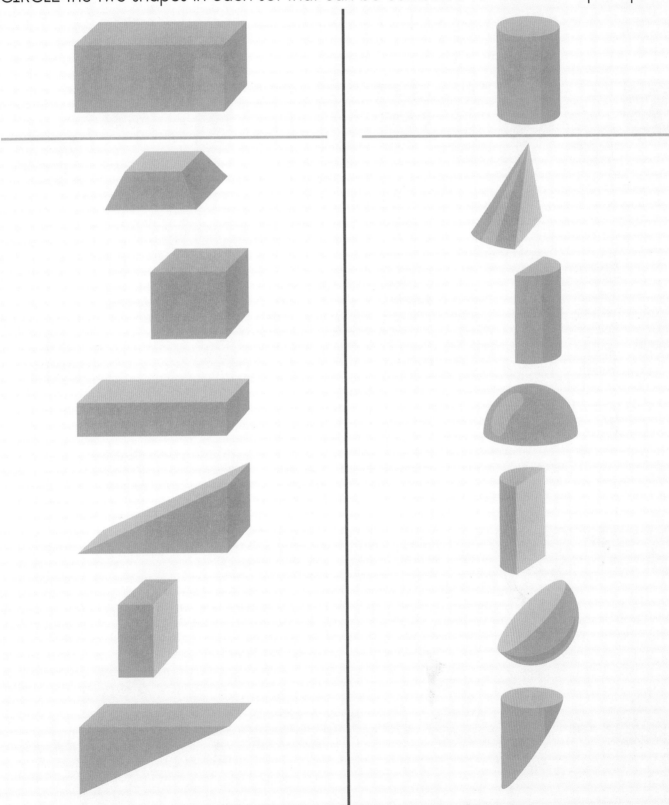

What Comes Next?

CIRCLE the shape that comes next in the pattern.

1.

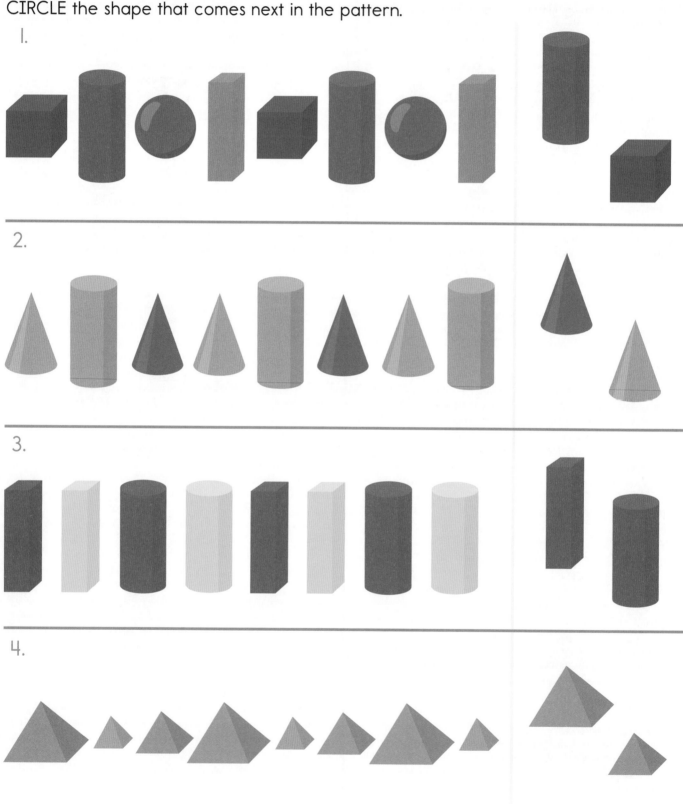

2.

3.

4.

Shapely Sequence

DRAW and COLOR shapes to finish each pattern.

Spiraling Sequence

DRAW and COLOR shapes to finish each pattern. Can you finish the spiral to the center?

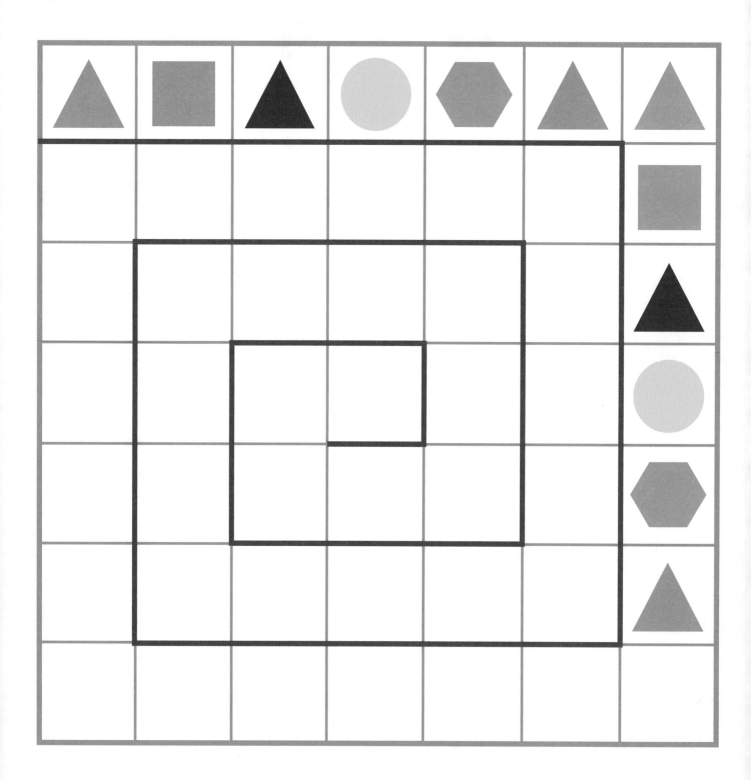

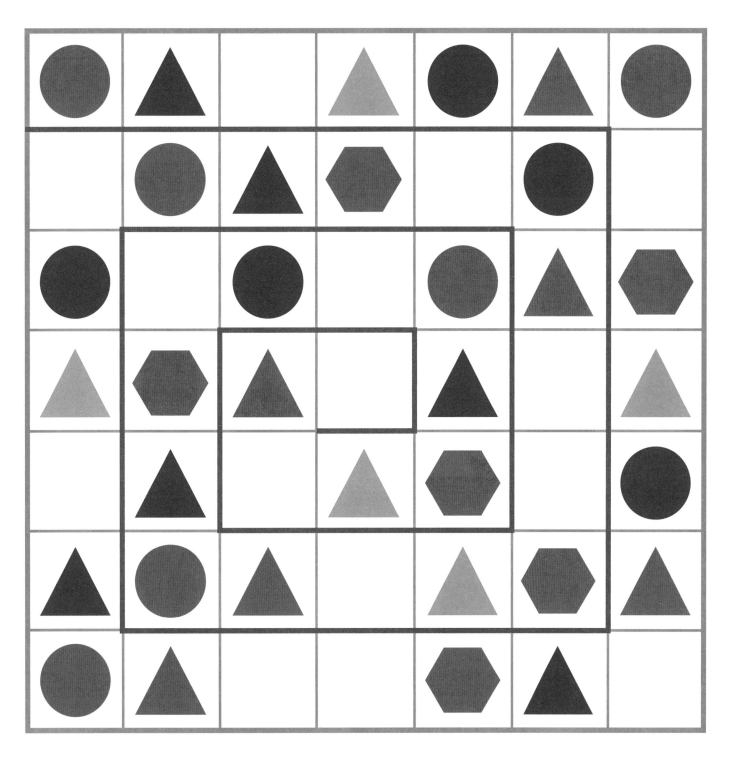

Shape Patterns

Puzzling Patterns

DRAW a line from each set of shapes to the place where it belongs in the pattern.

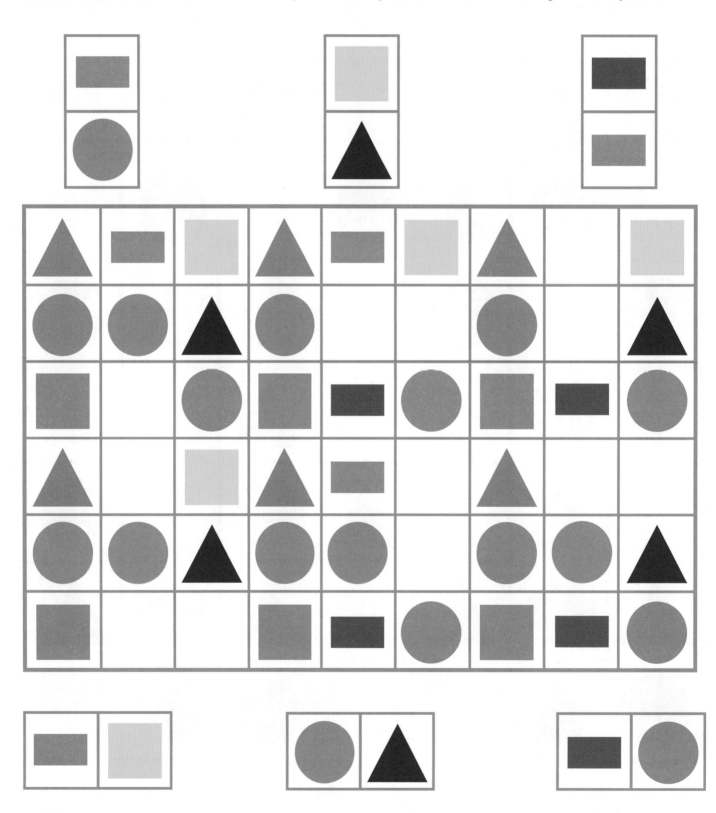

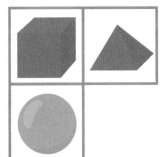

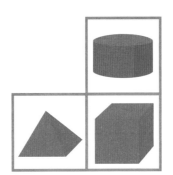

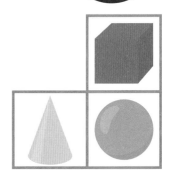

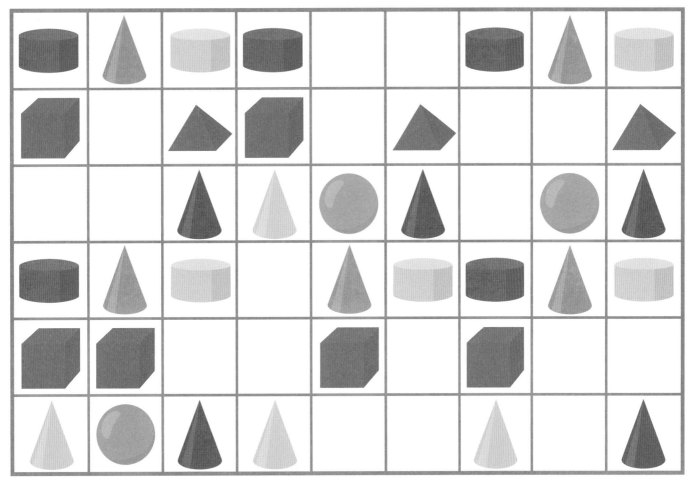

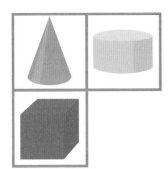

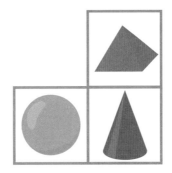

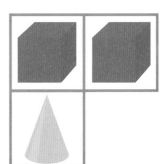

Matched Set

CUT OUT the cards on the opposite page. Using the blue cards, PLACE two cards next to each picture to make a matched set. (Save the cards to use again.)

HINT: Look for things the pictures have in common, like shape, color, or anything else! See how many different sets you can make.

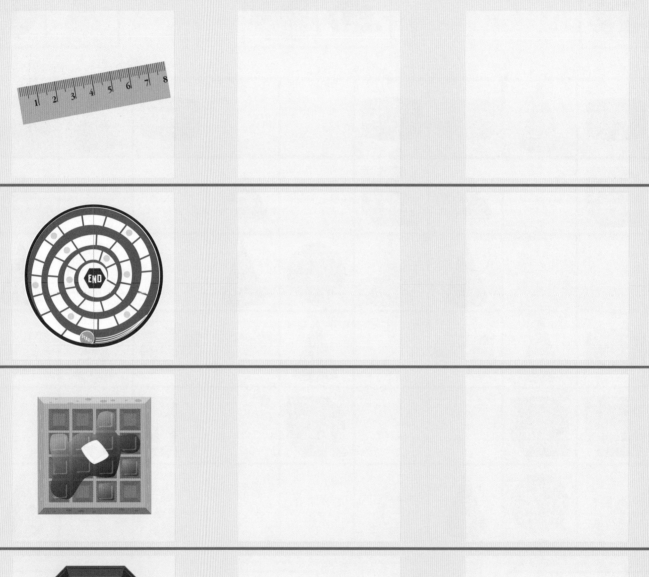

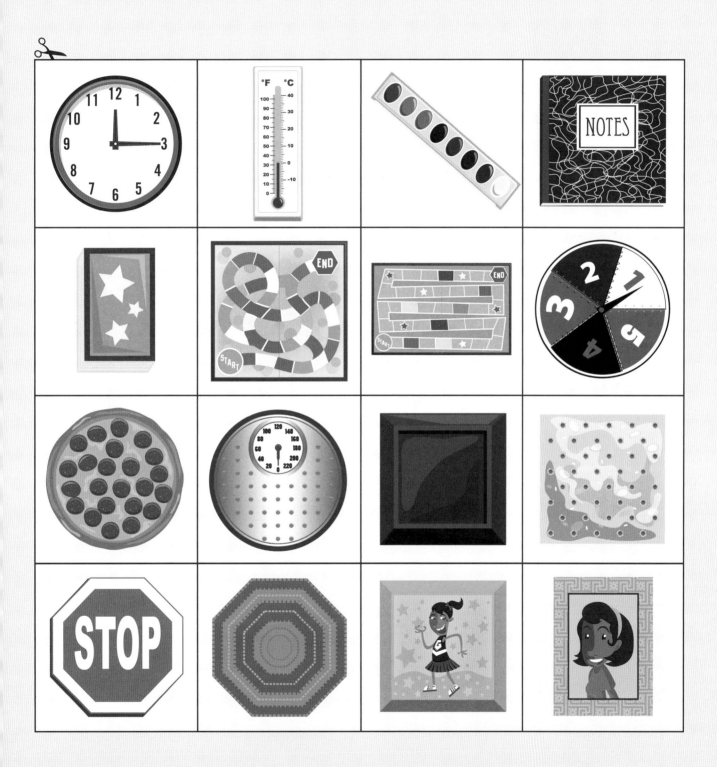

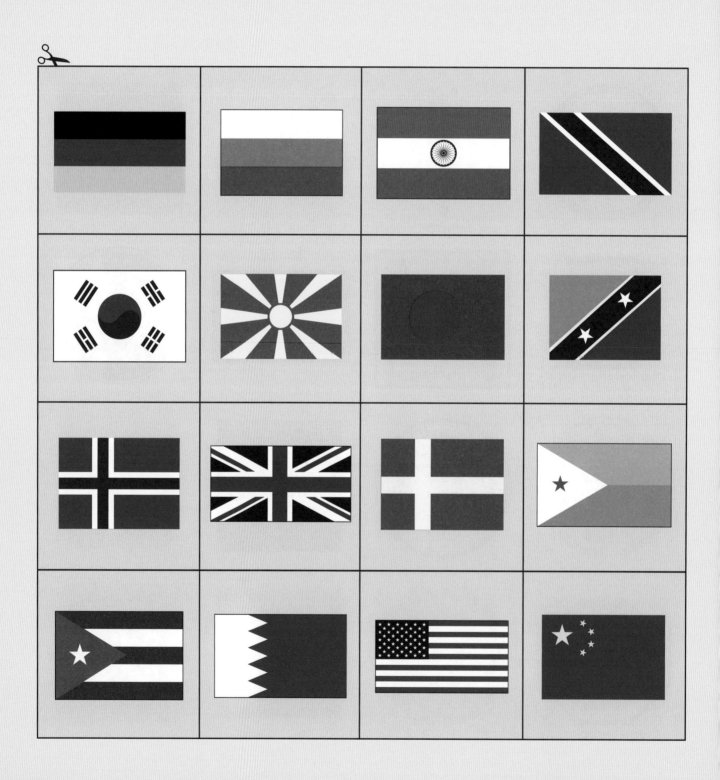

Matched Set

Using the purple cards from page 42, PLACE two cards next to each flag to make a matched set. (Save the cards to use again.)

HINT: Look for things the designs of the flags have in common, like stripes, circles, colors, or anything else! See how many different sets you can make.

Cross Out

CROSS OUT the shape in each set that does **not** have the same shape as the other shapes.

Squared Away

DRAW and COLOR the missing shapes.

Throw It Away

DRAW a line to put each piece of trash into the recycling container with the same shape.

HINT: Look at the shape of the trash and the shape of the recycling container.

Mad Lab

DRAW a line from each container on the counter to the shelf on the wall where it belongs.

Sorting Shapes

Get in Place

DRAW a line from each shape outside each diagram to show where it belongs inside the diagram.

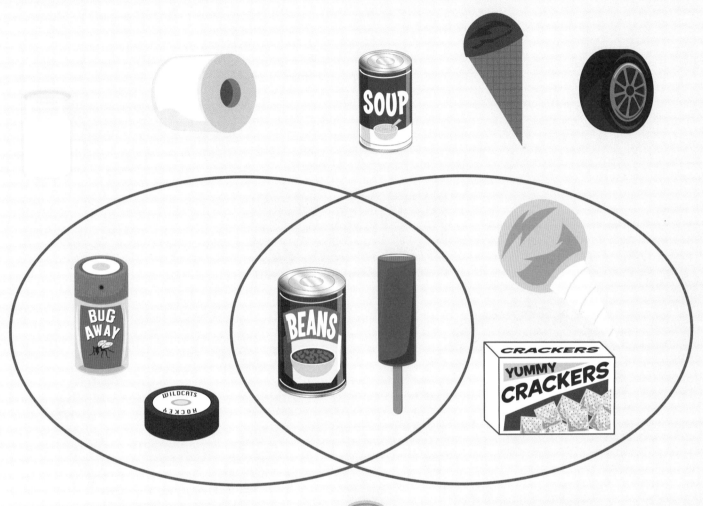

Perfect Plates

DRAW lines from the food to the plate with the same shape. CROSS OUT any food that does not belong on either plate.

Robot Repair

DRAW a line from each robot to its missing parts. CROSS OUT any parts that do not belong to a robot.

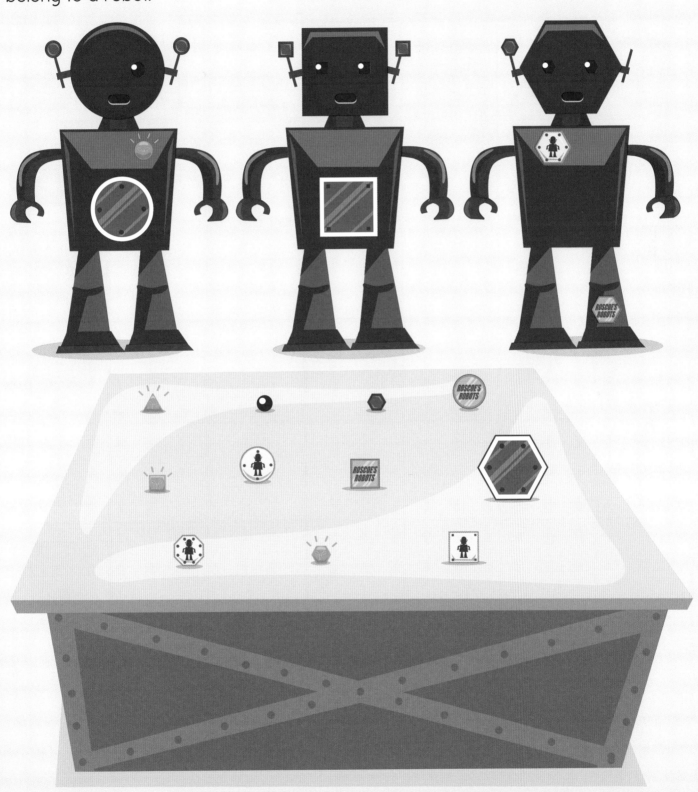

Review

Spiraling Sequence

DRAW and COLOR shapes to finish the pattern.

Puzzling Patterns

DRAW a line from each set of shapes to the place where it belongs in the pattern.

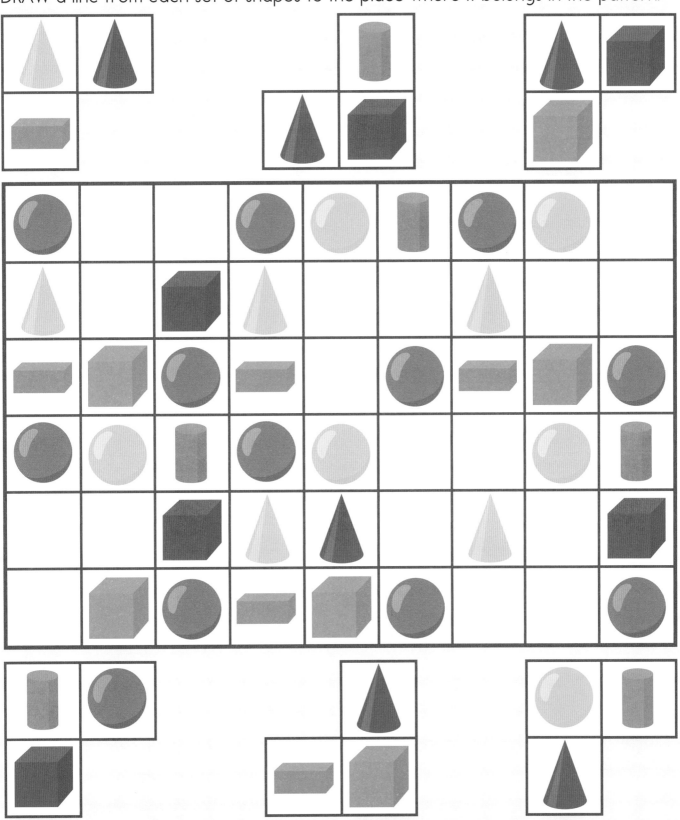

Matched Set

Using the purple cards from page 42, PLACE two cards next to each flag to make a matched set.

HINT: Look for things the designs of the flags have in common, like stripes, circles, colors, or anything else! See how many different sets you can make.

Get in Place

DRAW a line from each picture outside the diagram to show where it belongs inside the diagram.

Lines of Symmetry

Mirror, Mirror

A shape has **symmetry** if a line can divide the shape so each half is a mirror image of the other. DRAW a line of symmetry through each picture.

Line of symmetry

Symmetrical Superheroes

CIRCLE the superhero that is **not** symmetrical.

Lines of Symmetry

Mirror, Mirror

For each pair, FIND the picture that has symmetry. DRAW a line of symmetry for that picture.

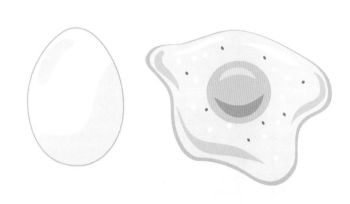

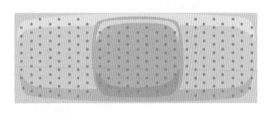

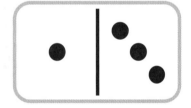

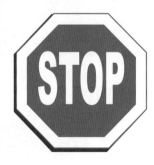

Line Counter

DRAW at least one line of symmetry for each shape. See how many lines of symmetry you can find.

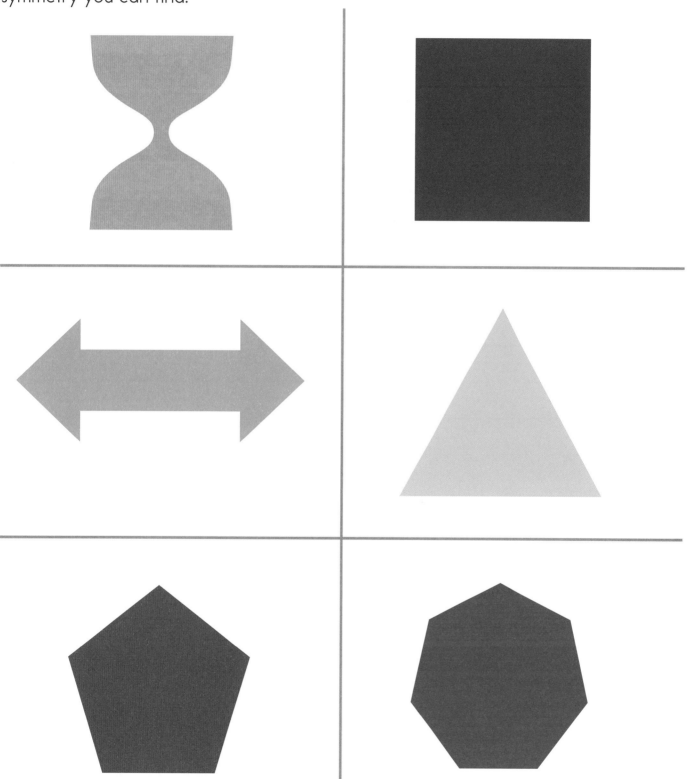

Color Flip

COLOR the pictures so they are symmetrical.

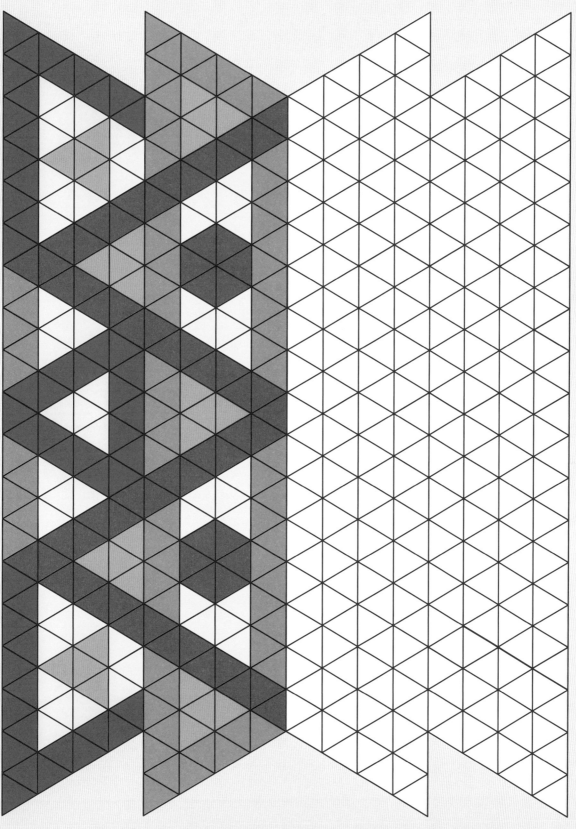

Shape Up

DRAW the mirror image of each shape, making it symmetrical. On the last two squares, DRAW your own symmetrical shapes.

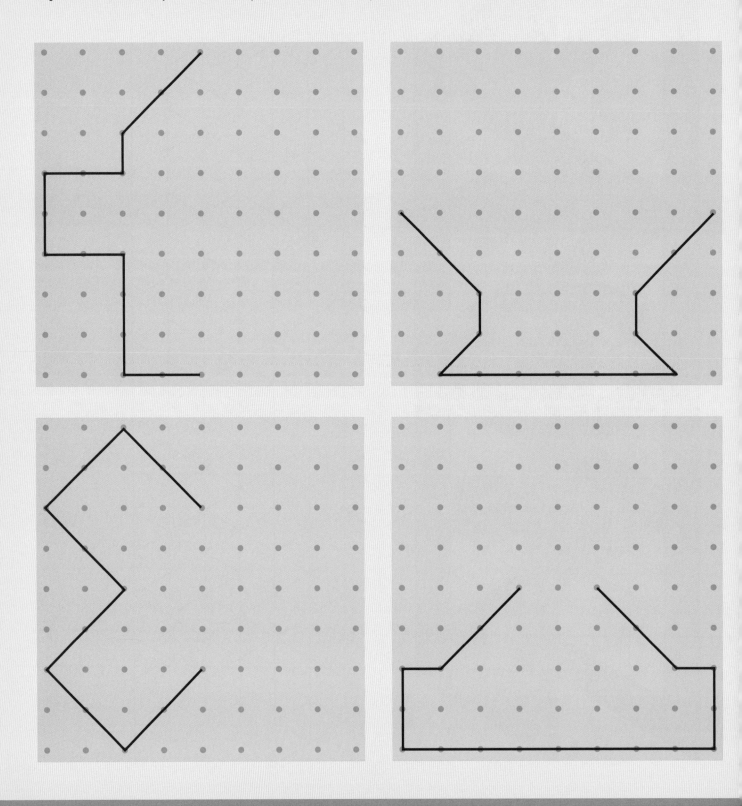

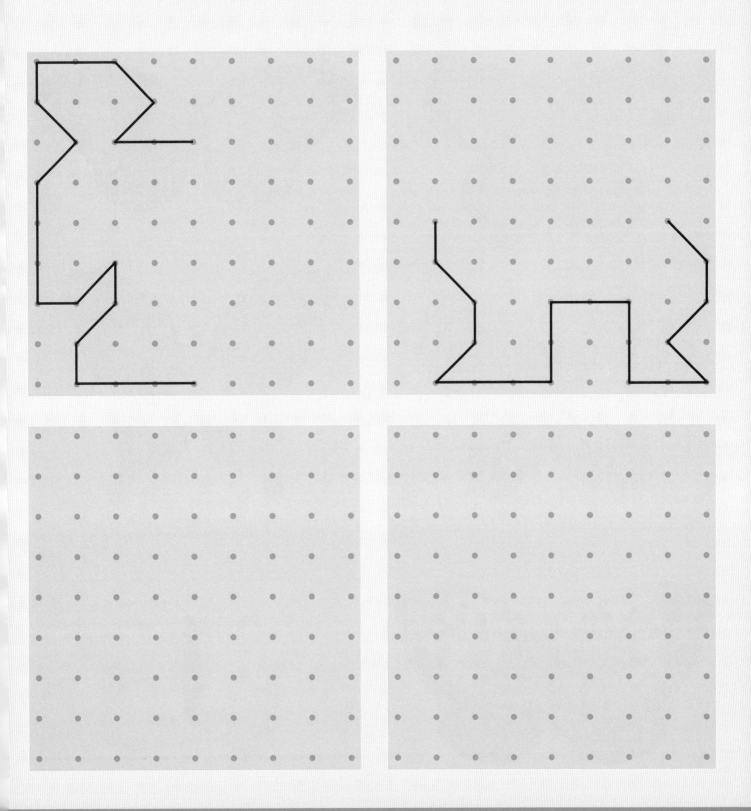

Beautiful Butterflies

CIRCLE the butterfly that is **not** symmetrical.

Mirror, Mirror

DRAW a line of symmetry through each letter.

HINT: Some letters have more than one line of symmetry.

A	D
T	O
E	M
H	I

Color Flip

COLOR the picture so it is symmetrical.

Shape Up

DRAW the mirror image of each shape, making it symmetrical.

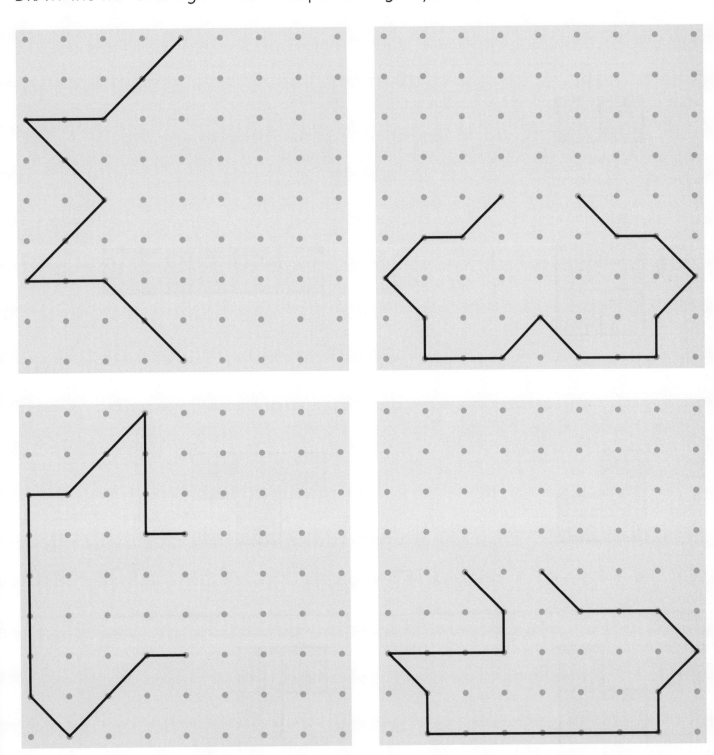

Perimeter

Squared Away

Perimeter is the distance around a two-dimensional shape. WRITE the perimeter of each shape.

Example: To measure the perimeter, count the number of units around the shape.

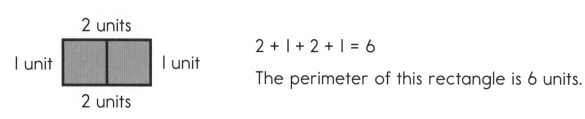

2 + 1 + 2 + 1 = 6

The perimeter of this rectangle is 6 units.

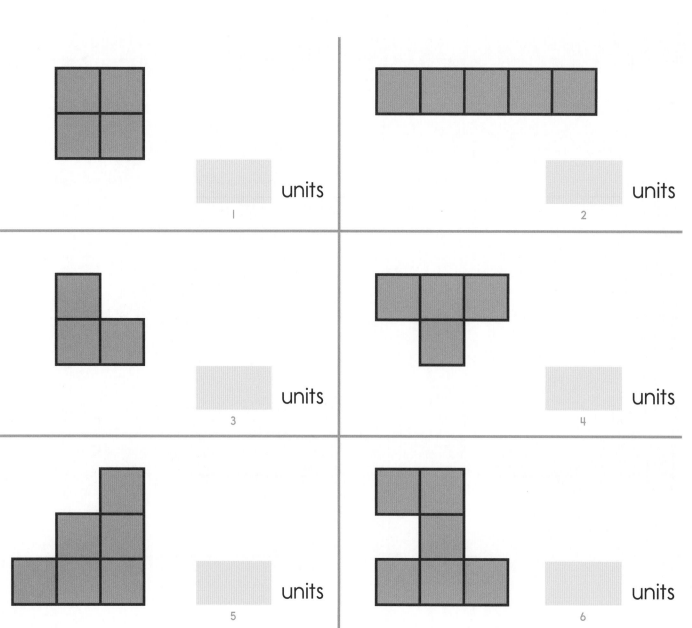

_____ units

1

_____ units

2

_____ units

3

_____ units

4

_____ units

5

_____ units

6

Puzzling Pentominoes

Using the pentomino pieces from page III, PLACE the pieces to completely fill each shape without overlapping any pieces. Then WRITE the perimeter of each shape. (Save the pieces to use later in the workbook.)

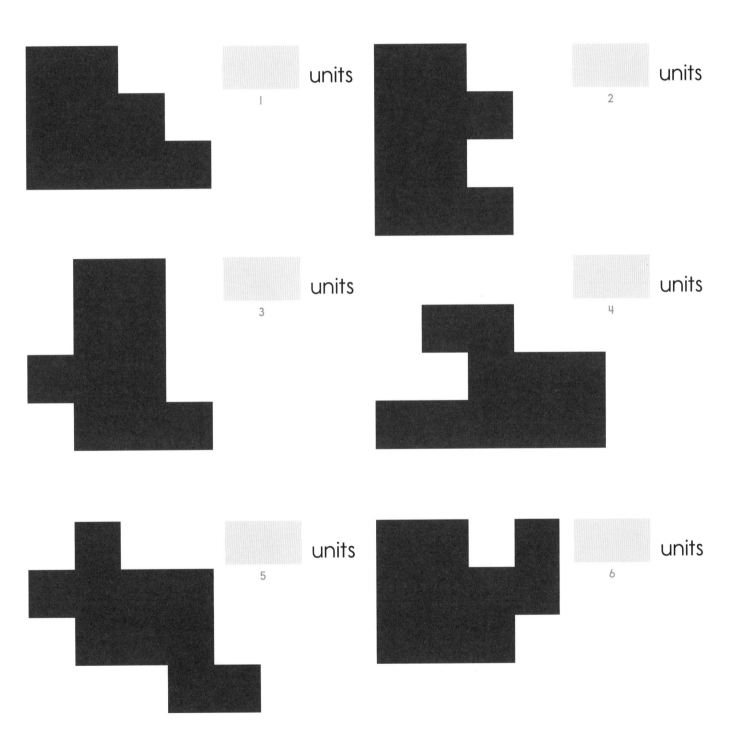

_____ units

1

_____ units

2

_____ units

3

_____ units

4

_____ units

5

_____ units

6

Perimeter

Around We Go

WRITE the perimeter of each shape.

Example:

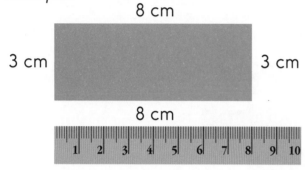

To find the perimeter, add the length of all of the sides.

8 + 3 + 8 + 3 = 22

The perimeter of this rectangle is 22 centimeters.

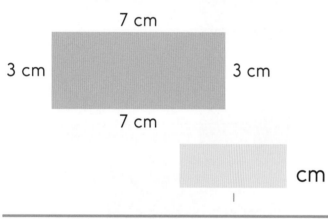

1

_____ cm

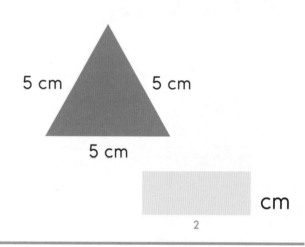

2

_____ cm

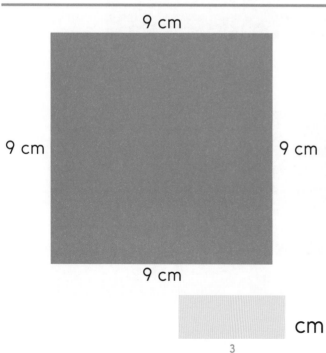

3

_____ cm

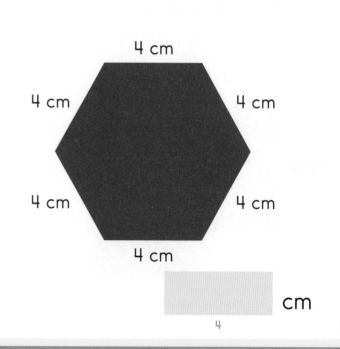

4

_____ cm

12

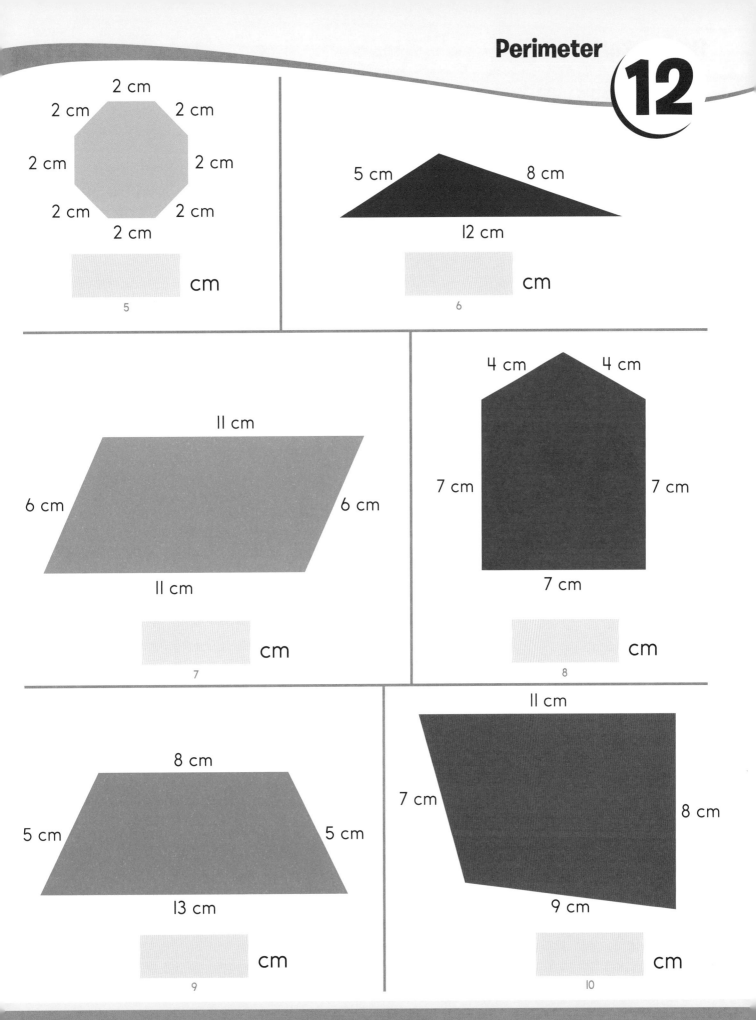

2 cm

2 cm   2 cm

2 cm   2 cm

2 cm   2 cm

2 cm

_____ cm

5

5 cm   8 cm

12 cm

_____ cm

6

11 cm

6 cm   6 cm

11 cm

_____ cm

7

4 cm   4 cm

7 cm   7 cm

7 cm

_____ cm

8

8 cm

5 cm   5 cm

13 cm

_____ cm

9

11 cm

7 cm   8 cm

9 cm

_____ cm

10

Perimeter

Farm Fences

Farmer Green wants to put fences around all of the animal areas of his farm. First he needs to know the perimeter of the animal areas so he knows how much fencing material to buy. WRITE the perimeter of each animal area.

Pigs: _____ ft
1

Cows: _____ ft
2

Chickens: _____ ft
3

Sheep: _____ ft
4

12

Shape Creator

DRAW four different shapes that all have a perimeter of 20 units.

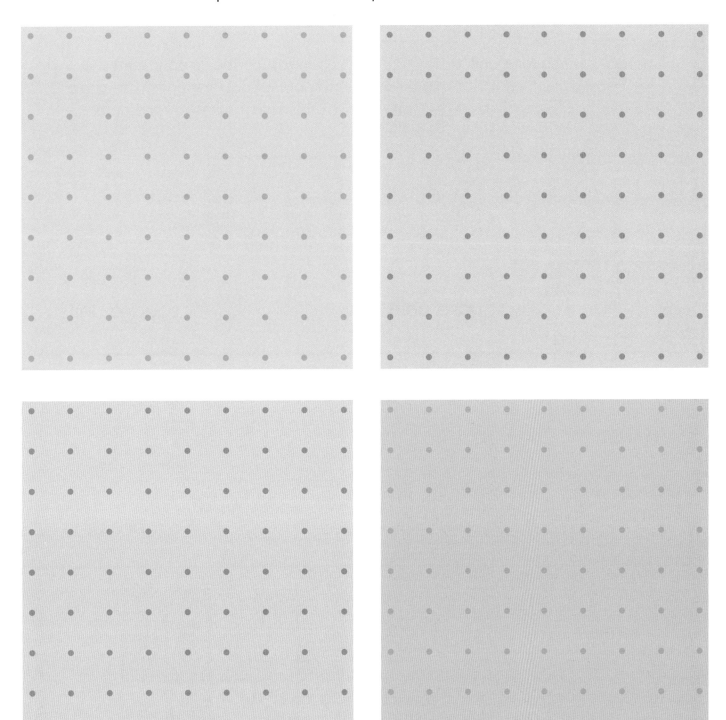

Squared Away

Area is the size of the surface of a shape, and it is measured in square units. WRITE the area of each shape.

Example:

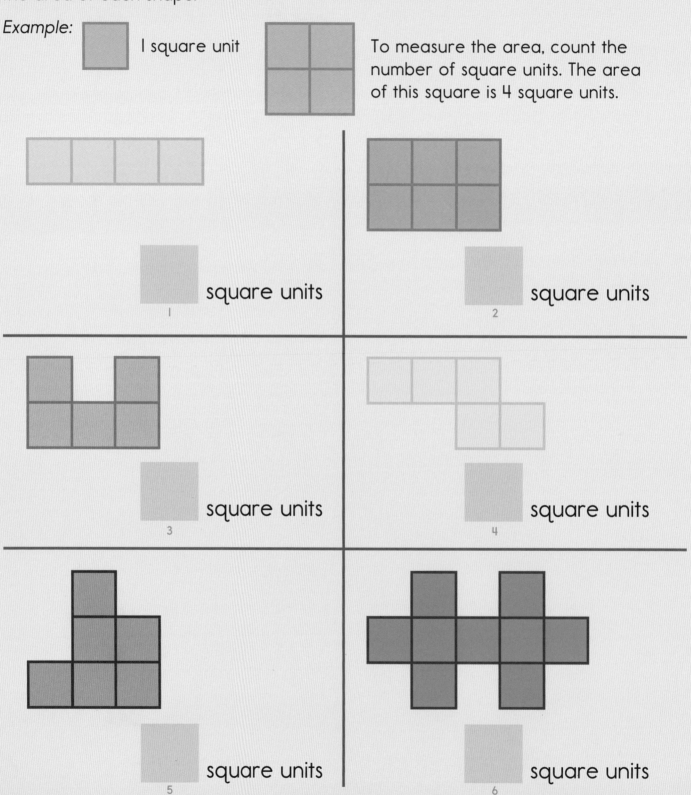

1 square unit

To measure the area, count the number of square units. The area of this square is 4 square units.

_____ square units
1

_____ square units
2

_____ square units
3

_____ square units
4

_____ square units
5

_____ square units
6

13

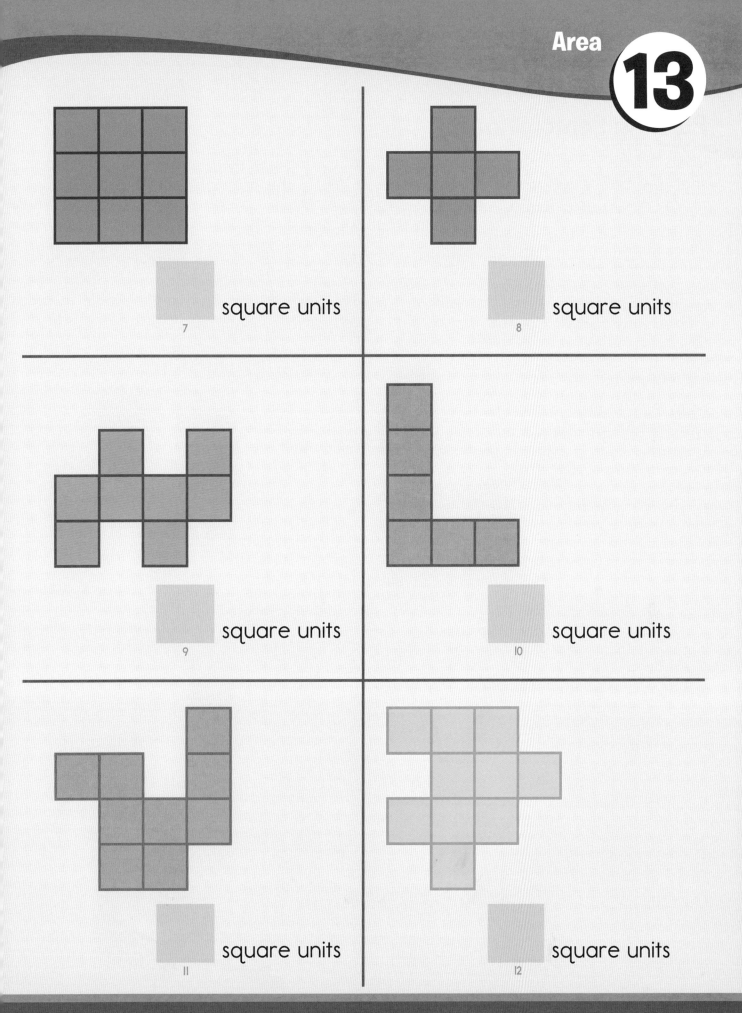

_____ square units

7

_____ square units

8

_____ square units

9

_____ square units

10

_____ square units

11

_____ square units

12

Area

Puzzling Pentominoes

Using the pentomino pieces from page III, PLACE the pieces to completely fill each shape without overlapping any pieces. Then WRITE the area of each shape. (Save the pieces to use later in the workbook.)

_____ square units

1

_____ square units

2

_____ square units

3

_____ square units

4

Tile Trouble

Samantha has four rooms in her house that need to be tiled, but she only has enough tiles for a room that is 22 square units. CIRCLE the rooms that she could choose to tile.

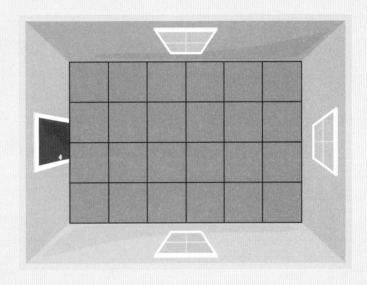

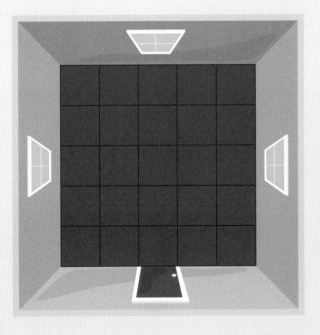

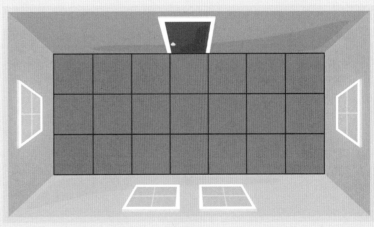

Square Pairs

DRAW lines to connect shapes that have the same area.

Shape Creator

DRAW four different shapes that all have an area of 16 square units.

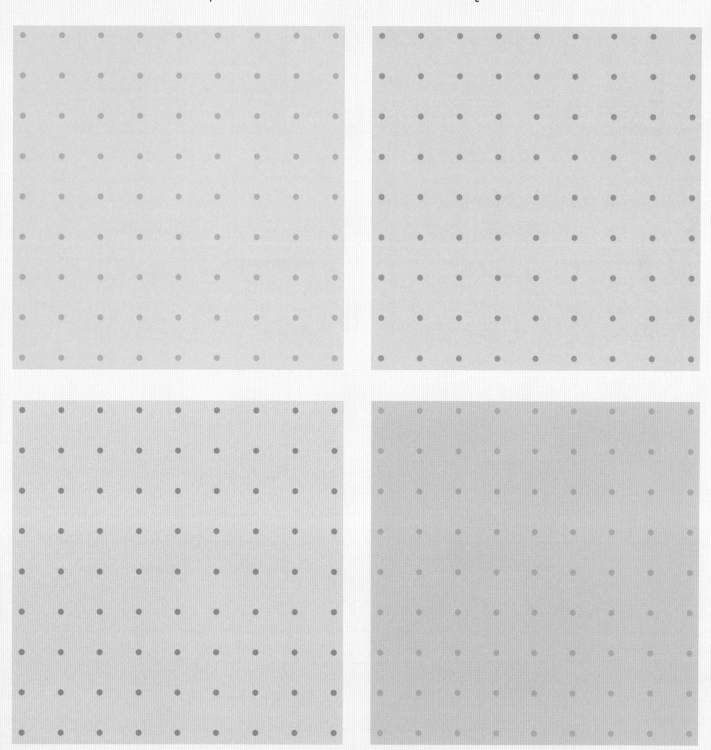

Squared Away

WRITE the perimeter and area of each shape.

1. Perimeter: units

   Area: square units

2. Perimeter: units

   Area: square units

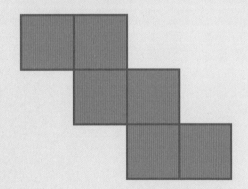

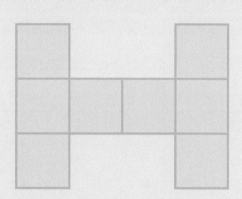

3. Perimeter: 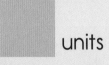 units

   Area: square units

4. Perimeter: units

   Area: square units

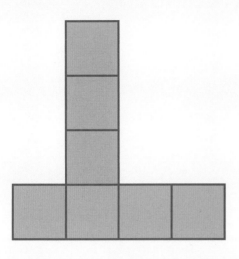

5. Perimeter: 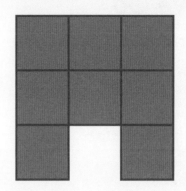 units

   Area:  square units

6. Perimeter: units

   Area: square units

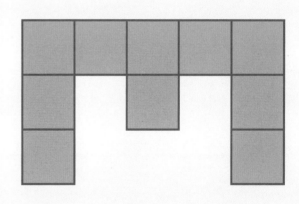

7. Perimeter: 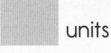 units

   Area: square units

8. Perimeter: units

   Area: square units

Around We Go

WRITE the perimeter of each shape.

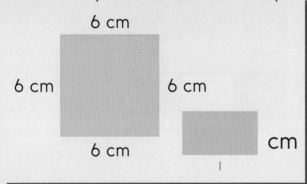

6 cm

6 cm   6 cm

6 cm

|____| cm

1

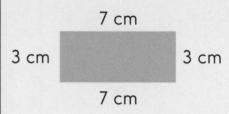

7 cm

3 cm   3 cm

7 cm

|____| cm

2

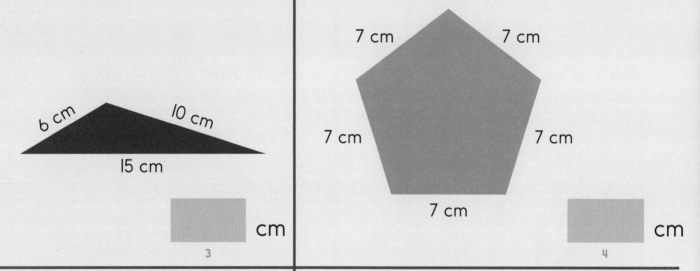

6 cm   10 cm

15 cm

|____| cm

3

7 cm   7 cm

7 cm   7 cm

7 cm

|____| cm

4

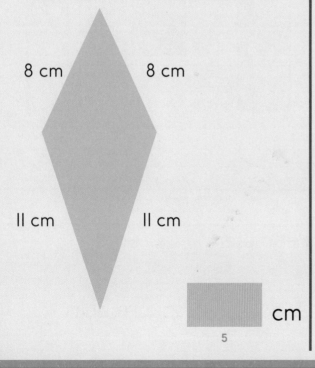

8 cm   8 cm

11 cm   11 cm

|____| cm

5

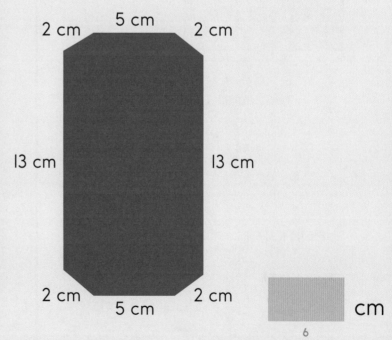

2 cm   5 cm   2 cm

13 cm   13 cm

2 cm   5 cm   2 cm

|____| cm

6

Dance Floor

Robin and Roscoe are looking for a
light-up dance floor for their dance
party. All of the dance floors have the
same area, so they're looking for one
with the largest perimeter. CIRCLE
the dance floor they should get.

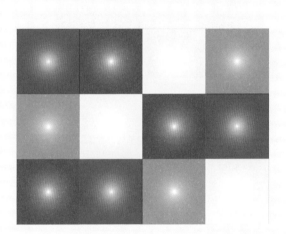

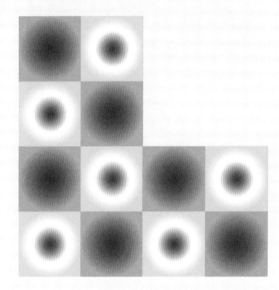

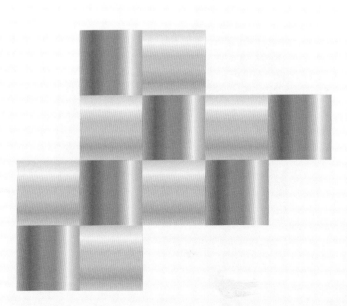

Puzzling Pentominoes

Using the pentomino pieces from page III, PLACE the pieces to completely fill each shape without overlapping any pieces. (Save the pieces to use again.)

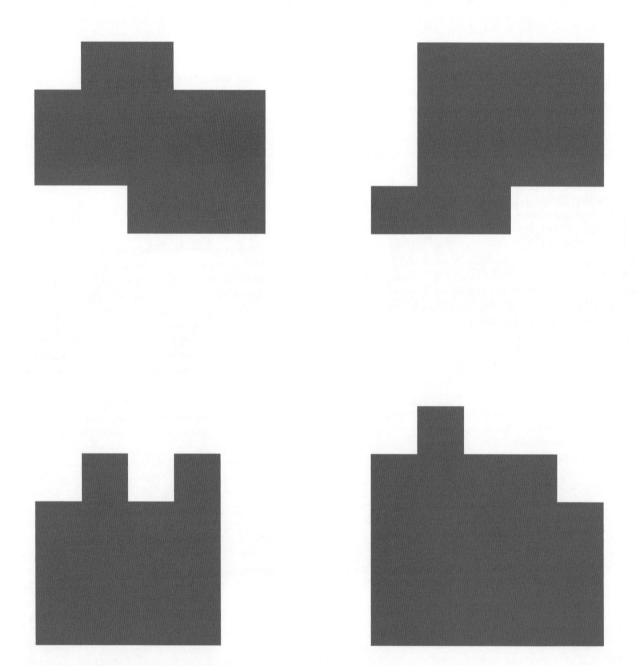

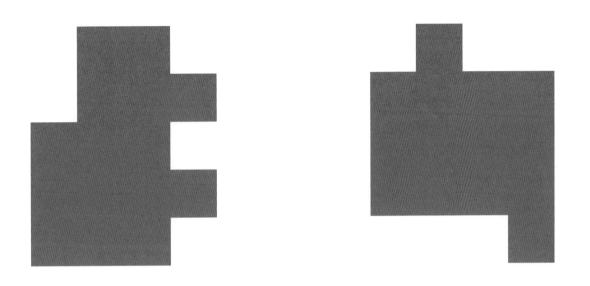

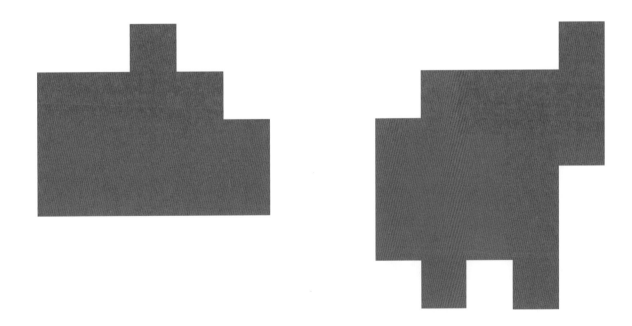

Puzzling Pentominoes

Using the pentomino pieces from page III, PLACE the pieces to completely fill each shape without overlapping any pieces. (Save the pieces to use again.)

Pentomino Place

Can you be the last player to make a move? Use all of the pentomino pieces from page III except the square. READ the rules. PLAY the game.

Rules: Two players

1. Give six pentomino pieces to each player.

2. Take turns placing pentomino pieces on the board without overlapping another piece.

The last player to be able to place a pentomino piece wins!

HINT: Try to place your pieces to block the other player.

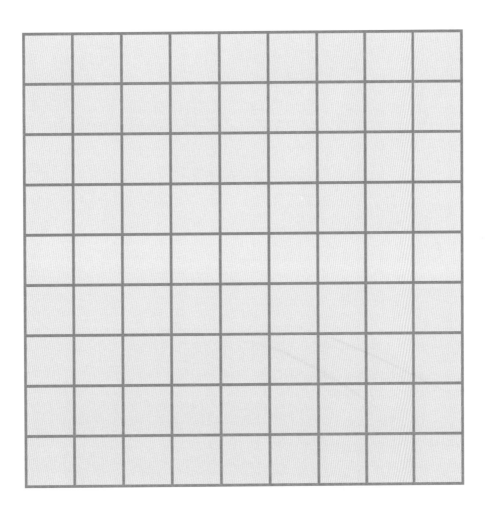

Shape Shifters

Using the pattern block pieces from page 113, PLACE the pieces to completely fill each shape without overlapping any pieces. See if you can solve the puzzles in different ways. (Save the pieces to use again.)

Symmetry Shapes

Using the pattern block pieces from page 113, PLACE the pieces to make each picture symmetrical without overlapping any pieces. (Save the pieces to use again.)

Pattern Blocks

15

Tricky Tangrams

Using the tangram pieces from page 115, PLACE the pieces to completely fill each shape without overlapping any pieces. (Save the pieces to use again.)

HINT: Try placing the biggest pieces first. You do not need to use all of the tangram pieces in each shape.

Symmetry Shapes

Using the tangram pieces from page II5, PLACE the pieces to make each picture symmetrical without overlapping any pieces. (Save the pieces to use again.)

Puzzling Pentominoes

Using the pentomino pieces from page III, PLACE the pieces to completely fill each shape without overlapping any pieces.

Shape Shifters

Using the pattern block pieces from page 113, PLACE the pieces to completely fill each shape without overlapping any pieces. See if you can solve the puzzles in different ways. (Save the pieces to use again.)

Symmetry Shapes

Using the pattern block pieces from page 113, PLACE the pieces to make the picture symmetrical without overlapping any pieces.

Tricky Tangrams

Using the tangram pieces from page 115, PLACE the pieces to completely fill each shape without overlapping any pieces.

Shapes Squared

FOLLOW the directions in order, and DRAW and COLOR the shapes in the correct squares on the opposite page.

1. Draw a green triangle below the purple hexagon.

2. Draw a yellow rectangle between the blue triangle and orange circle.

3. Draw an orange square to the right of the red rectangle.

4. Draw a red circle above the purple hexagon.

5. Draw a blue hexagon under the orange circle.

6. Draw a green square to the left of the blue triangle.

7. Draw a purple triangle over the orange square.

8. Draw a red circle to the right of the green triangle.

9. Draw a blue square below the red rectangle.

10. Draw a yellow triangle between the blue hexagon and the red circle.

11. Draw an orange triangle to the left of the purple triangle.

12. Draw a yellow circle between the blue square and the green triangle.

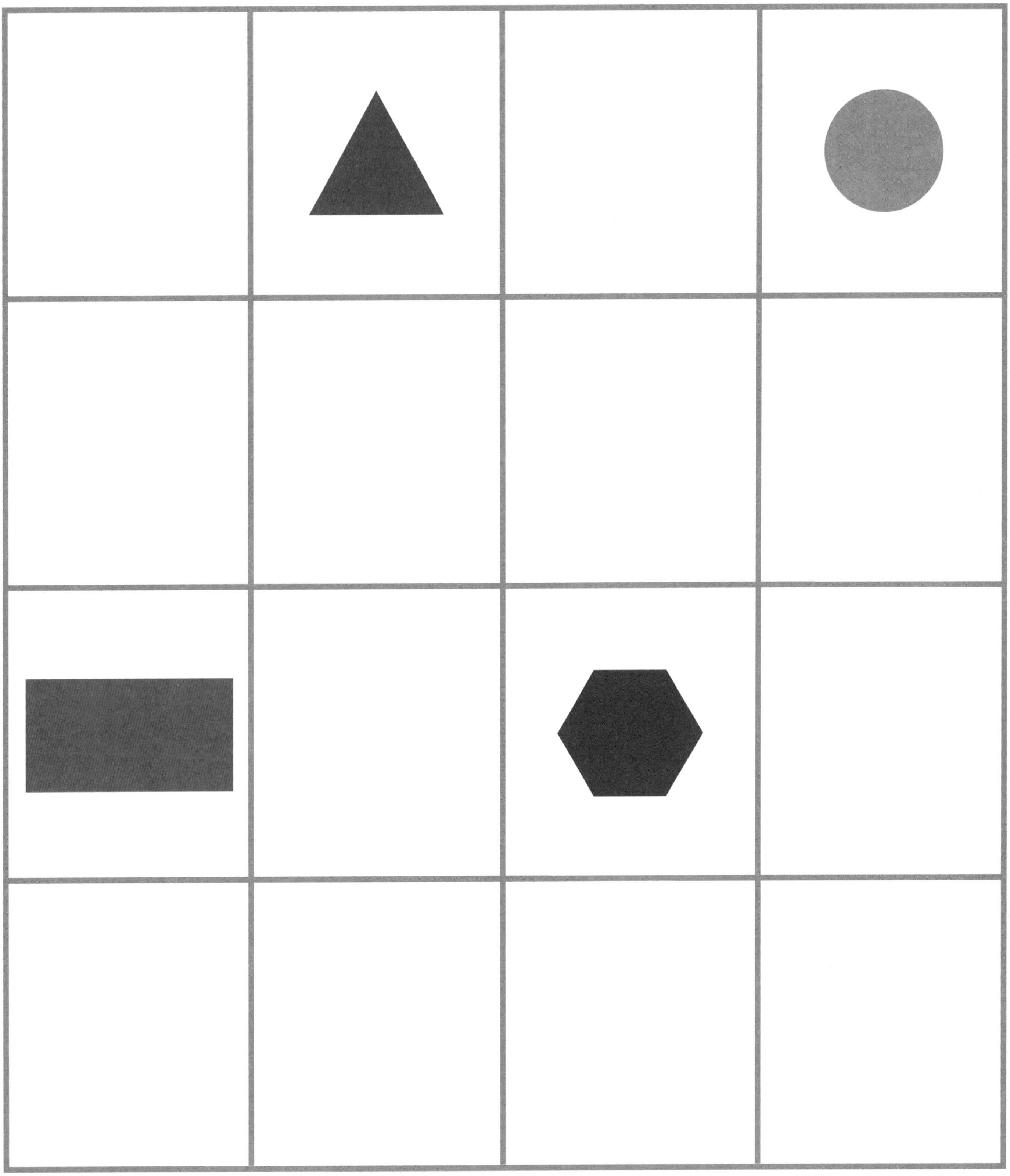

Who Am I?

LOOK at the picture on the opposite page. Then CIRCLE the correct person.

1. I am standing above the pond.

2. I am sitting on a bench far from the fountain.

3. I am standing behind a hot dog cart.

4. I am standing next to the fountain but not near the bridge.

5. I am standing on the grass to the right of the pond.

6. I am sitting on a bench next to the hot dog cart but not in front of the pond.

Maps

Who Am I?

FOLLOW the directions, and CIRCLE the correct location of each person.

1.  I live in the yellow house. I left my house and went past the bakery. I turned left on Chestnut Street and then right on Maple Street. I went two blocks and turned left on Hazelnut Street. I went to the end of the block. Where am I?

2.  I live in the red house on the corner of Maple and Walnut Street. I went along Maple Street and turned left to go down Olive Street. I turned left again when I got to Oak Street but stopped before I crossed Chestnut Street again. Where am I?

3.  I live in the blue house. I went down Fig Street and turned left on Willow Street, riding past my favorite restaurant. I turned right on Chestnut Street and then quickly turned left on Maple Street. When I got to Walnut Street, I turned right and went a little more than a block. Where am I?

4. I live in the green house. The first thing I did when I left my house is stop by the diner on my block. From the diner I went up Hazelnut Street and turned left on Oak Street. I passed the police station and the department store but turned right on Chestnut Street before I reached the shoe store. I did not make any more turns, and I did not cross Maple Street. Where am I?

Pine Street

Willow Street

Maple Street

Fig Street

Olive Street

Chestnut Street

Walnut Street

Hazelnut Street

Oak Street

Elm Street

Maps

Castle Quest

The knight has been called to duty at the castle. FOLLOW the directions, and DRAW the path to his castle.

HINT: Look at the compass to find north, south, east, and west.

To the castle:
Go east three spaces.
Go north five spaces.
Go west one space.
Go north two spaces.
Go east four spaces.
Go south three spaces.
Go east two spaces.
Go south five spaces.
Go east two spaces.

Go for a Ride

WRITE the name of the ride that can be found at each location on the map.

HINT: Follow the letter and the number, and see where the two lines meet.

1. A3 _____

2. D6 _____

3. C5 _____

4. A1 _____

5. E4 _____

6. B7 _____

7. D1 _____

8. C3 _____

9. E7 _____

10. A5 _____

Shapes Squared

FOLLOW the directions in order, and DRAW and COLOR the shapes in the correct squares.

1. Draw a yellow circle to the right of the purple octagon.

2. Draw a red square below the green hexagon.

3. Draw an orange triangle above the yellow circle.

4. Draw a green circle to the left of the blue rectangle.

5. Draw a blue square between the green hexagon and the orange triangle.

6. Draw an orange hexagon under the red square.

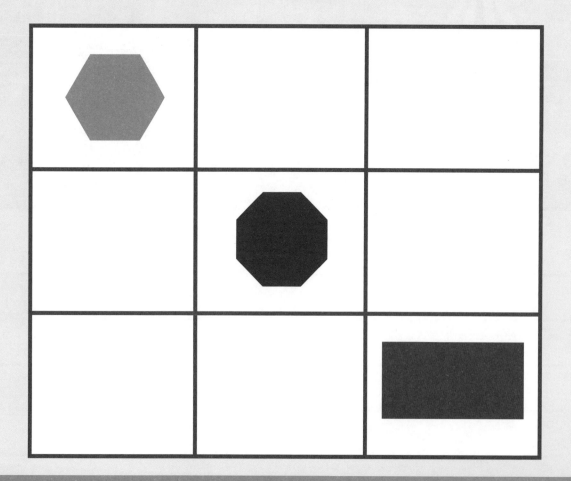

Castle Quest

The knight is trapped in a castle, and the princess must rescue him! FOLLOW the directions, and DRAW the path to his castle.

To the castle:
Go west three spaces.
Go north one space.
Go west three spaces.
Go south three spaces.
Go east two spaces.
Go south six spaces.
Go west two spaces.
Go south one space.
Go west two spaces.
Go north four spaces.
Go west two spaces.

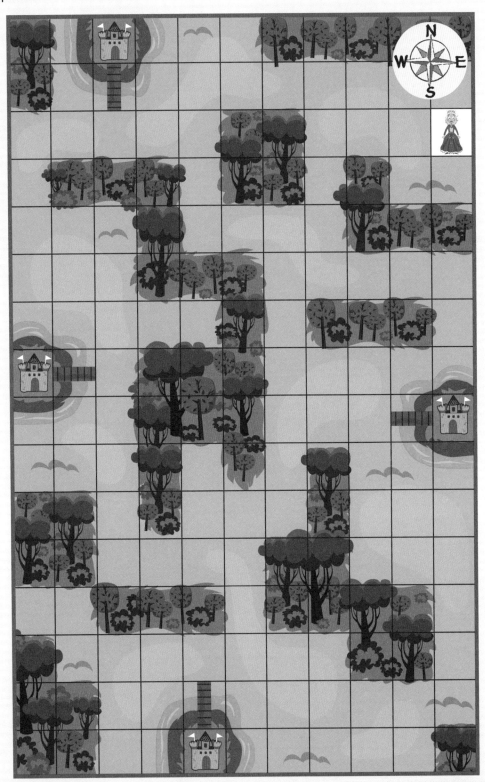

Awesome Aquarium

WRITE the name of the animal that can be found at each location on the map.

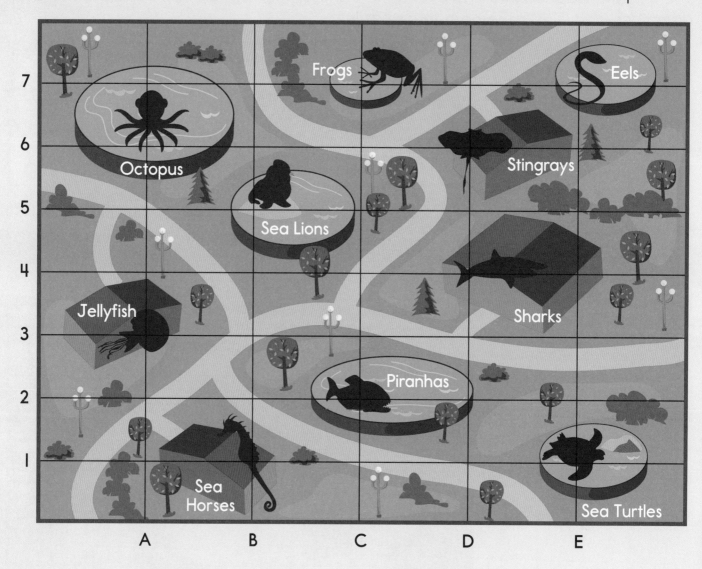

1. A6 _____

2. D4 _____

3. C2 _____

4. B1 _____

5. E1 _____

6. D6 _____

7. B5 _____

8. E7 _____

9. A3 _____

10. C7 _____

Pentominoes

CUT OUT the 13 pentomino pieces.

These pentomino pieces are for use with pages 69, 76, 84, 85, 86, 87, and 96.

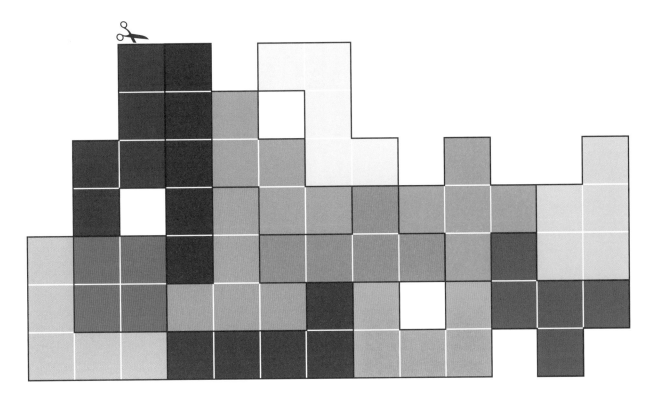

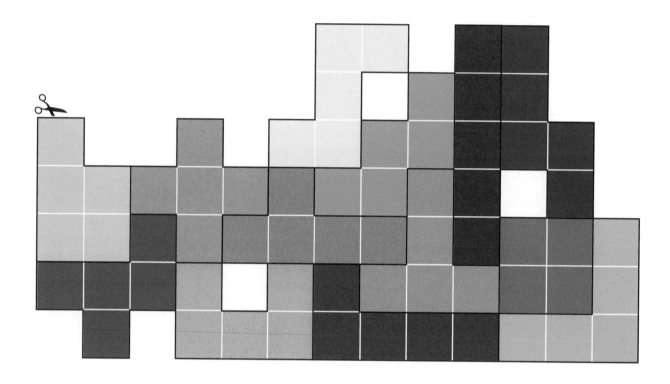

Pattern Blocks

CUT OUT the 31 pattern block pieces.

These pattern block pieces are for use with pages 88, 89, 90, 91, 97, and 98.

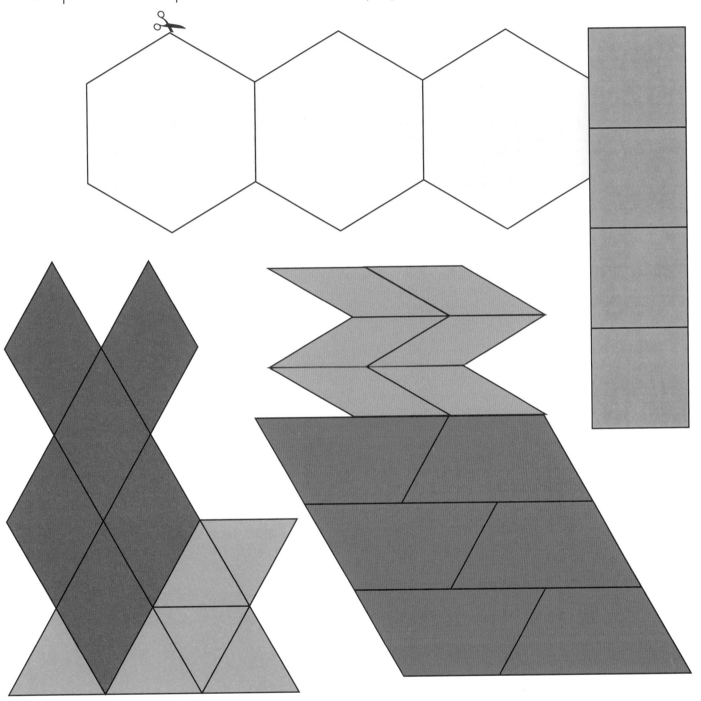

Game Pieces

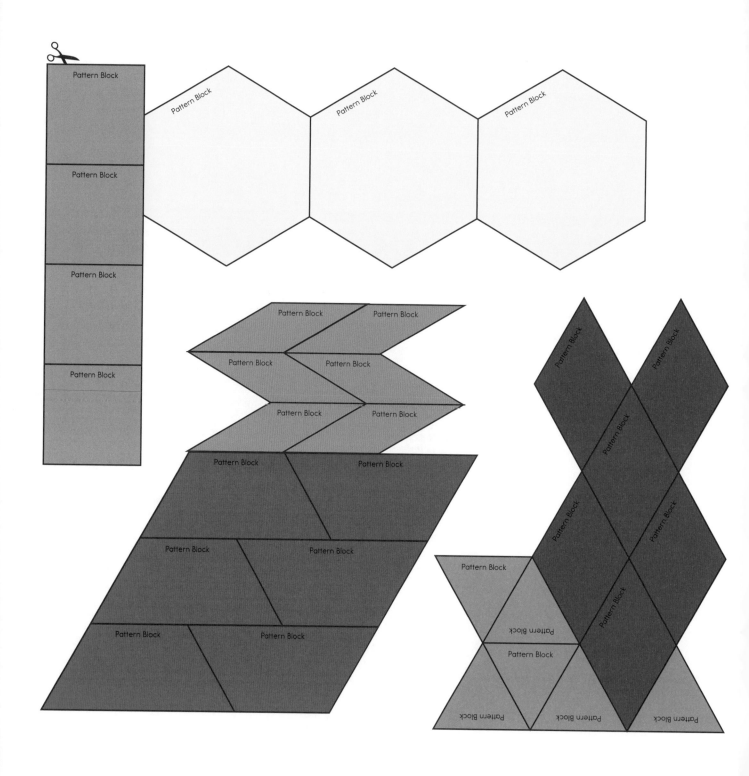

Tangrams

CUT OUT the seven tangram pieces.

These tangram pieces are for use with pages 92, 93, 94, 95, and 99.

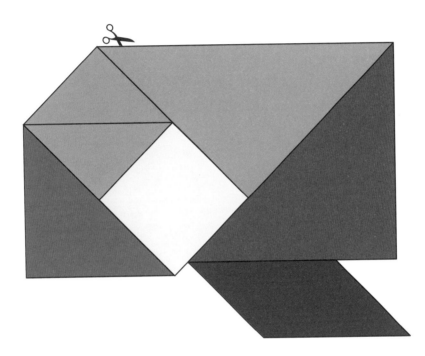

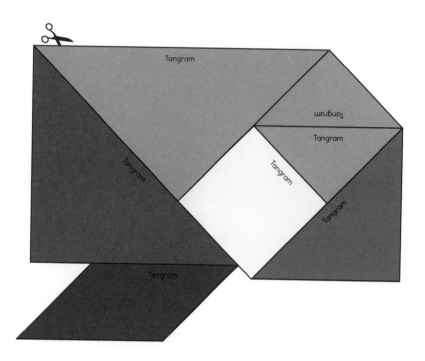

Answers

Page 2
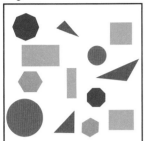

Page 3

Page 4
1. 1
2. 3
3. 4
4. 2
5. 3

Page 5
Have someone check your answers.

Page 6
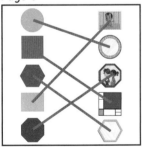

Page 7
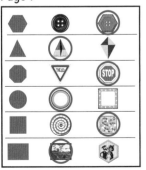

Page 8

Page 9
Have someone check your answers.

Page 10
Suggestion:

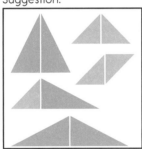

Page 13
1. 5
2. 9
3. 14
4. 13

Page 14
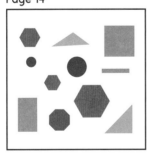

Page 15
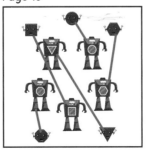

Page 16
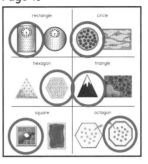

Page 17
Suggestion:

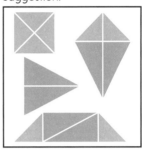

Page 18
1. cylinder
2. rectangular prism
3. square pyramid
4. cone
5. sphere
6. rectangular prism
7. cube
8. cylinder
9. square pyramid

Page 19

Page 20
1. 2
2. 3
3. 5
4. 1
5. 2

Page 21
Have someone check your answers.

Page 22
1. cube
2. rectangular prism
3. – 6. Have someone check your answers.

Page 25

Page 26
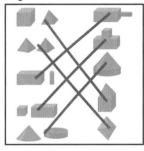

Page 27

Page 28
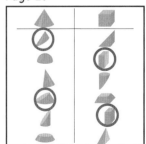

Answers

Page 29

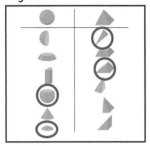

Page 30
1. rectangular prism
2. cone
3. sphere
4. cube
5. square pyramid
6. cylinder
7. sphere
8. square pyramid
9. cube
10. rectangular prism
11. cylinder
12. cone

Page 31

Page 32

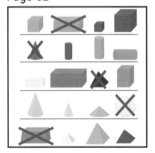

Page 33

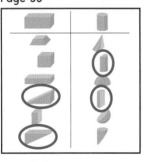

Page 34

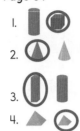

Page 35

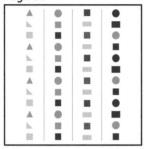

Page 36

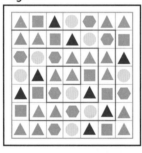

Page 37

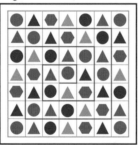

Page 38

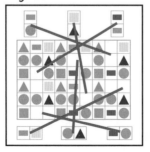

Page 39

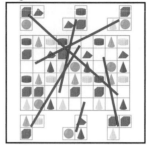

Page 40
Suggestion:

Page 43
Suggestion:

Page 44

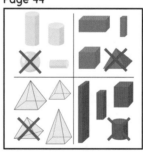

Page 45

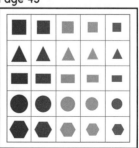

Page 46

Page 47

Page 48

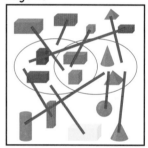

Page 49

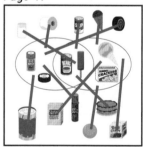

Page 50

Page 51

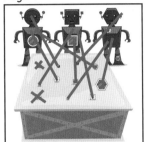

Page 52

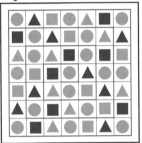

Page 53

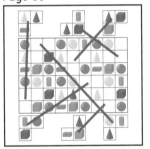

Page 54
Suggestion:

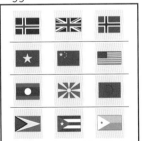

Page 55

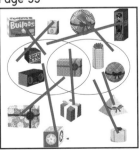

Page 56

Page 57

Page 58
Suggestion:

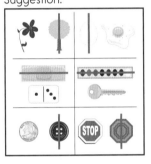

Page 59

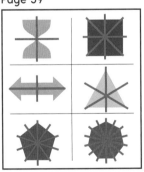

Page 60

Page 61

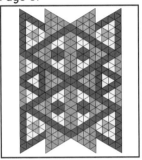

Page 62

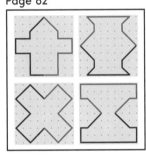

Page 63

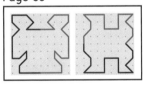

Have someone check
your answers.

Page 64

Page 65

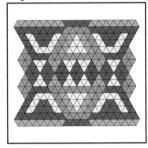

Page 66

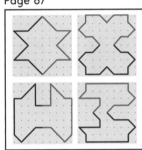

Page 67

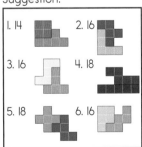

Page 68
1. 8   2. 12
3. 8   4. 10
5. 12   6. 14

Page 69
Suggestion:

1. 14   2. 16

3. 16   4. 18

5. 18   6. 16

Answers

Pages 70–71

1. 20	2. 15
3. 36	4. 24
5. 16	6. 25
7. 34	8. 29
9. 31	10. 35

Page 72

1. 40	2. 70
3. 38	4. 60

Page 73

Have someone check your answers.

Pages 74–75

1. 4	2. 6
3. 5	4. 5
5. 6	6. 9
7. 9	8. 5
9. 8	10. 6
11. 9	12. 10

Page 76

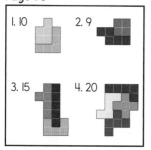

Page 77

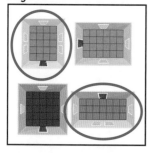

Page 78

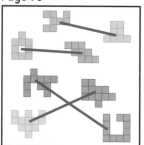

Page 79

Have someone check your answers.

Pages 80–81

1. 10, 4	2. 12, 8
3. 14, 6	4. 18, 8
5. 16, 7	6. 14, 8
7. 16, 8	8. 22, 10

Page 82

1. 24	2. 20
3. 31	4. 35
5. 38	6. 44

Page 83

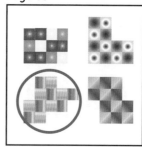

Page 84
Suggestion:

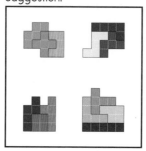

Page 85
Suggestion:

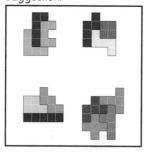

Page 86
Suggestion:

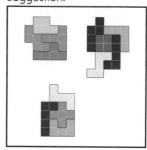

Page 88
Suggestion:

Page 89
Suggestion:

Page 90

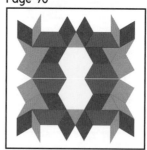

Page 91

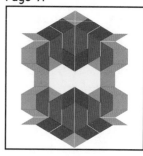

Page 92
Suggestion:

Page 93
Suggestion:

Page 94

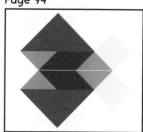

Page 95

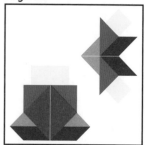

Answers

Page 96
Suggestion:

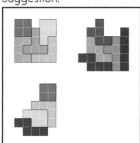

Page 97
Suggestion:

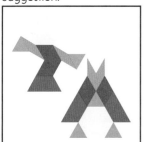

Page 98

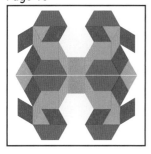

Page 99
Suggestion:

Pages 100–101

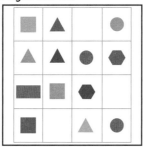

Pages 102–103

1.

2.

3.

4.

5.

6.

Pages 104–105

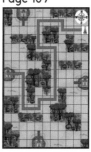

Page 106

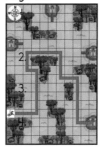

Page 107
1. Carousel
2. Bumper Cars
3. Super Spinner
4. Teacups
5. Zipper
6. Logger's Run
7. Crazy Coaster
8. Ferris Wheel
9. Octorama
10. Sky Twirl

Page 108

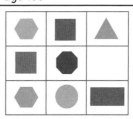

Page 109

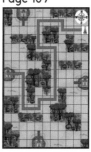

Page 110
1. Octopus
2. Sharks
3. Piranhas
4. Sea Horses
5. Sea Turtles
6. Stingrays
7. Sea Lions
8. Eels
9. Jellyfish
10. Frogs

Just a page per day with Sylvan workbooks helps kids catch up, keep up, and get ahead!

EXTRA PRACTICE
Just a
PAGE
PER DAY
THE EASY WAY

Review and Improve Skills • Grow Self-Confidence • Develop a Love of Learning!

Visit SylvanPagePerDay.com to find out more about doing a page per day of practice at home.

"If you are looking for some good, fun, learning books for your child, I definitely recommend the Sylvan Learning series."

—www.thedadjam.com

"**Samantha loves these books** because to her, they are not school work. They are fun activities. But really she is learning, and doing the same work she does at school."

—www.mommymandy.com

"As a teacher, I look for different aspects in a resource than a student or a parent might. . . . **The book has vibrant, attractive colors, and the worksheets are really fun.** My 10-year-old son has been enjoying them since I got the book. . . . I recommend this book to students, parents, and teachers alike for increasing student achievement."

—www.dynamitelessonplan.com

Sylvan workbooks help kids catch up for the next report card or get ahead for the next school year.

READING & LANGUAGE ARTS

Kindergarten Alphabet Activities (Workbook) $12.99/$15.50 Can.
Kindergarten Beginning Word Games (Workbook) $12.99/$15.50 Can.
Kindergarten Language Arts Success (Super Workbook) $18.99/$22.00 Can.
Kindergarten Reading Readiness (Workbook) $12.99/$15.50 Can.
Kindergarten Success with Sight Words (Workbook) $12.99/$14.99 Can. **NEW!**
Success with Lowercase Letters: Grades K–1 (Workbook) $12.99/$14.99 Can. **NEW!**
Success with Uppercase Letters: Grades K–1 (Workbook) $12.99/$14.99 Can. **NEW!**
First Grade Language Arts Success (Super Workbook) $18.99/$22.00 Can.
First Grade Reading Skill Builders (Workbook) $12.99/$15.50 Can.
First Grade Spelling Games & Activities (Workbook) $12.99/$15.50 Can.
First Grade Success with Sight Words (Workbook) $12.99/$14.99 Can. **NEW!**
First Grade Vocabulary Puzzles (Workbook) $12.99/$15.50 Can.
Second Grade Language Arts Success (Super Workbook) $18.99/$22.00 Can.
Second Grade Reading Skill Builders (Workbook) $12.99/$15.50 Can.
Second Grade Spelling Games & Activities (Workbook) $12.99/$15.50 Can.
Second Grade Success with Sight Words (Workbook) $12.99/$14.99 Can. **NEW!**
Second Grade Vocabulary Puzzles (Workbook) $12.99/$15.50 Can.
Third Grade Reading Comprehension Success (Workbook) $12.99/$15.00 Can.
Third Grade Reading Success: Complete Learning Kit $79.00/$92.00 Can.
Third Grade Spelling Success (Workbook) $12.99/$15.00 Can.
Third Grade Super Reading Success (Super Workbook) $18.99/$22.00 Can.
Third Grade Vocabulary Success (Workbook) $12.99/$15.00 Can.
Fourth Grade Reading Comprehension Success (Workbook) $12.99/$15.00 Can.
Fourth Grade Reading Success: Complete Learning Kit $79.00/$92.00 Can.
Fourth Grade Spelling Success (Workbook) $12.99/$15.00 Can.
Fourth Grade Super Reading Success (Super Workbook) $18.99/$22.00 Can.
Fourth Grade Vocabulary Success (Workbook) $12.99/$15.00 Can.
Fifth Grade Reading Comprehension Success (Workbook) $12.99/$15.00 Can.
Fifth Grade Reading Success: Complete Learning Kit $79.00/$92.00 Can.
Fifth Grade Super Reading Success (Super Workbook) $18.99/$22.00 Can.

Fifth Grade Vocabulary Success (Workbook) $12.99/$15.00 Can.
Fifth Grade Writing Success (Workbook) $12.99/$15.00 Can.

MATH

Pre-K Shapes & Measurement Success (Workbook) $12.99/$14.99 Can.
Kindergarten Basic Math Success (Workbook) $12.99/$15.99 Can.
Kindergarten Math Games & Puzzles (Workbook) $12.99/$15.99 Can.
Kindergarten Shapes & Geometry Success (Workbook) $12.99/$14.99 Can.
First Grade Basic Math Success (Workbook) $12.99/$15.99 Can.
First Grade Math Games & Puzzles (Workbook) $12.99/$15.99 Can.
First Grade Shapes & Geometry Success (Workbook) $12.99/$14.99 Can.
Second Grade Basic Math Success (Workbook) $12.99/$15.99 Can.
Second Grade Geometry Success (Workbook) $12.99/$14.99 Can.
Second Grade Math Games & Puzzles (Workbook) $12.99/$15.99 Can.
Second Grade Math in Action (Workbook) $12.99/$14.99 Can.
Second Grade Super Math Success (Super Workbook) $18.99/$21.99 Can.
Third Grade Basic Math Success (Workbook) $12.99/$15.99 Can.
Third Grade Geometry Success (Workbook) $12.99/$14.99 Can.
Third Grade Math Games & Puzzles (Workbook) $12.99/$15.99 Can.
Third Grade Math in Action (Workbook) $12.99/$14.99 Can.
Third Grade Super Math Success (Super Workbook) $18.99/$21.99 Can.
Fourth Grade Basic Math Success (Workbook) $12.99/$15.99 Can.
Fourth Grade Geometry Success (Workbook) $12.99/$14.99 Can.
Fourth Grade Math Games & Puzzles (Workbook) $12.99/$15.99 Can.
Fourth Grade Math in Action (Workbook) $12.99/$14.99 Can.
Fourth Grade Super Math Success (Super Workbook) $18.99/$21.99 Can.
Fifth Grade Basic Math Success (Workbook) $12.99/$15.99 Can.
Fifth Grade Math Games & Puzzles (Workbook) $12.99/$15.99 Can.
Fifth Grade Geometry Success (Workbook) $12.99/$14.99 Can.
Fifth Grade Math in Action (Workbook) $12.99/$14.99 Can.
Fifth Grade Super Math Success (Super Workbook) $18.99/$21.99 Can.